AF807673

Naughty nudity plagued my upbringing. I was not a Playboy model/centerfold. I wasn't the type. I didn't want to be. What was it about the naked female form that sold so many magazines? What was different from the "Photography Today" magazine's nudity? Was the 'high' art nudity a cover, a lie to allow pornography to exist, to allow what was an intrusive photographic "looking machine" to see with composition, and interpretation and, by implication, art? Was naked photography imitating high art nudity or was it just exposure of the vulgar and not expressive, just prurient voyeurism and definitively NOT art? I was having some issues with composition, wanting to be a three-dimensional artist, so I decided to classify, like Aristotle, the Female Form Frozen (mostly on film) in two-dimensional popular photography as practice for later 3d compositions. I got permission from Playboy, which was to occupy most of my classification efforts. Seven years later, when I had completed the collages for the book, Playboy in Sears Tower told me they had different lawyers and would not allow derivative works

This is a look at what AI thinks about the female form, despite using only free firefly images, compared to what photographers of stock images and others think of the female form.

AI LEGS

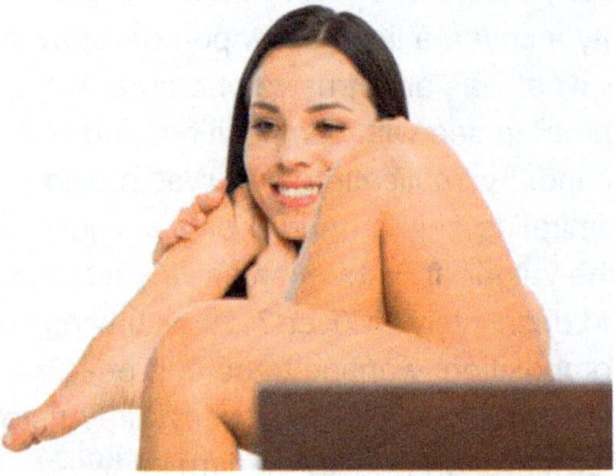

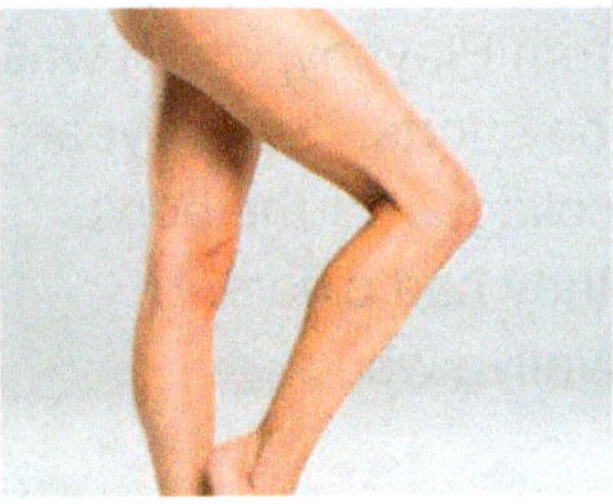

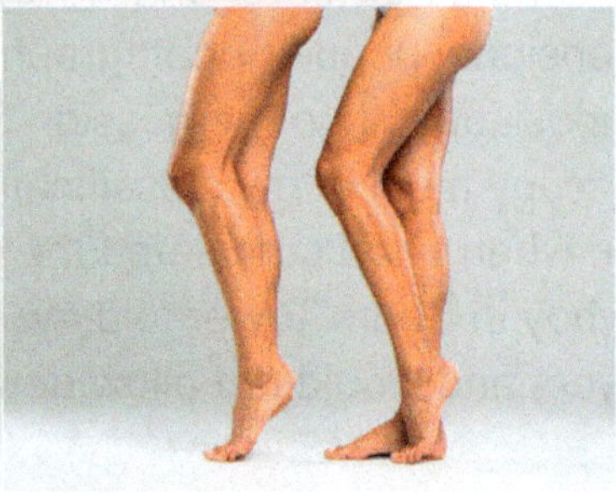

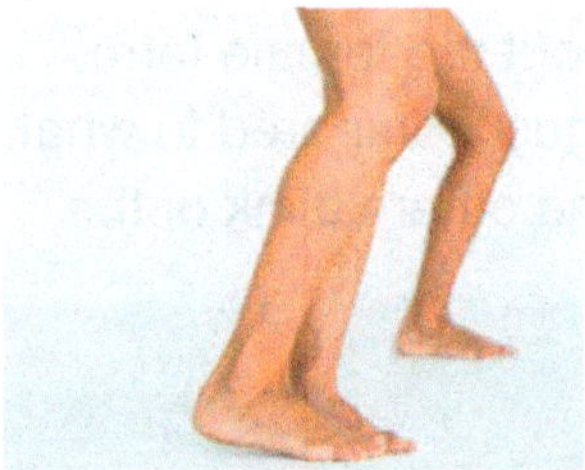

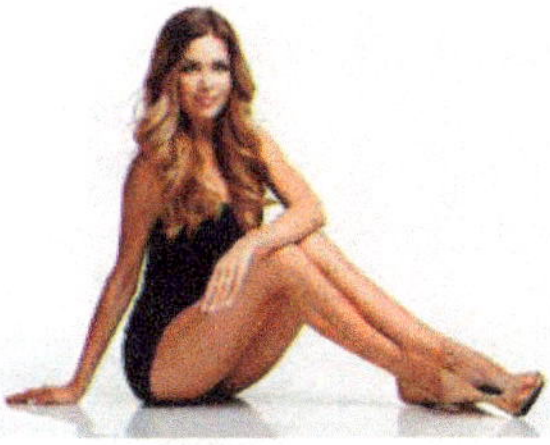

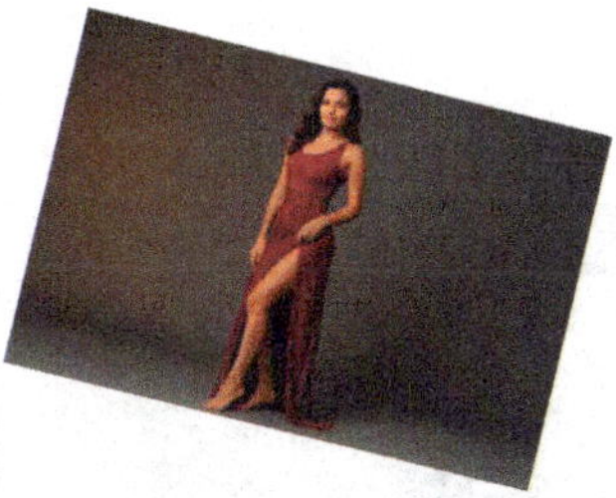

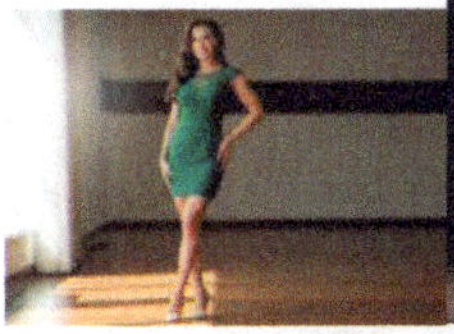

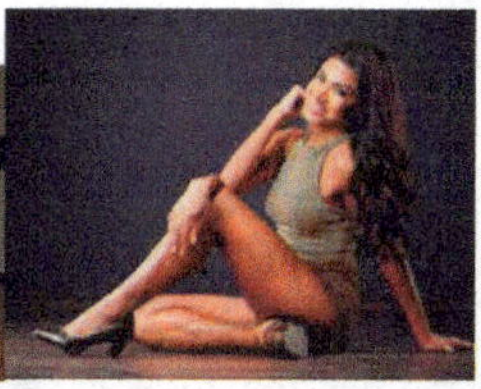

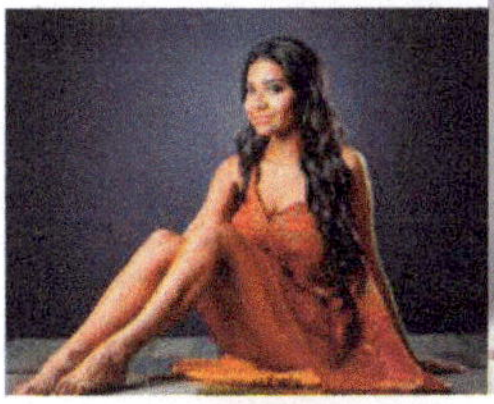

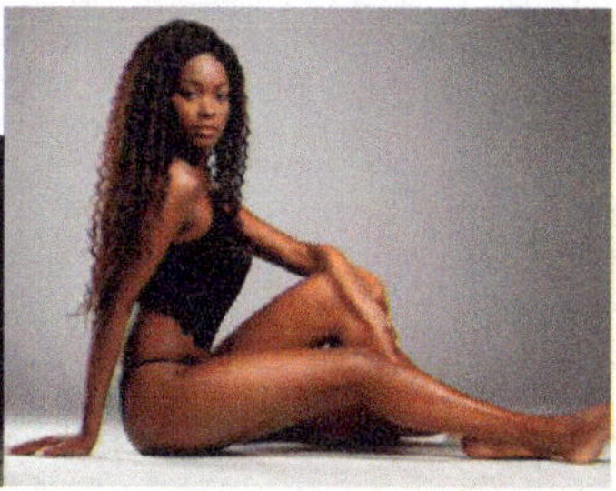

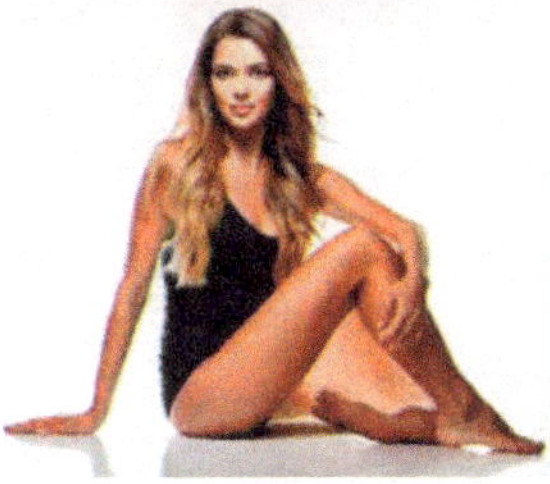

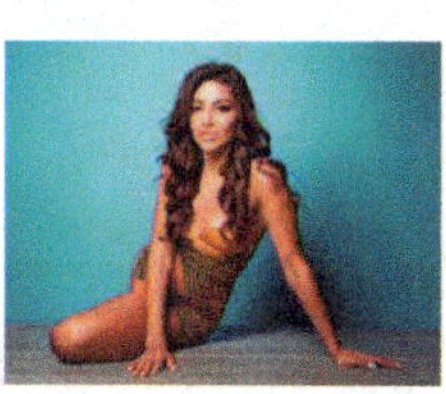

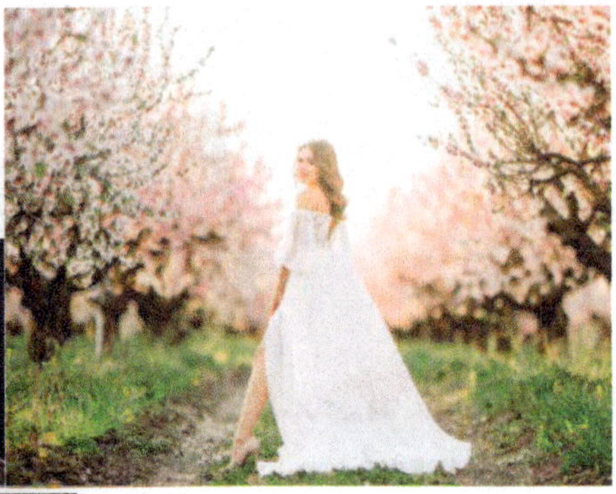
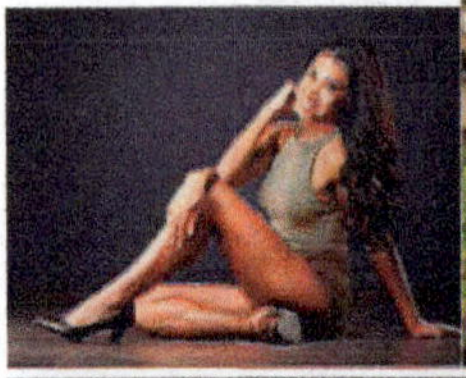
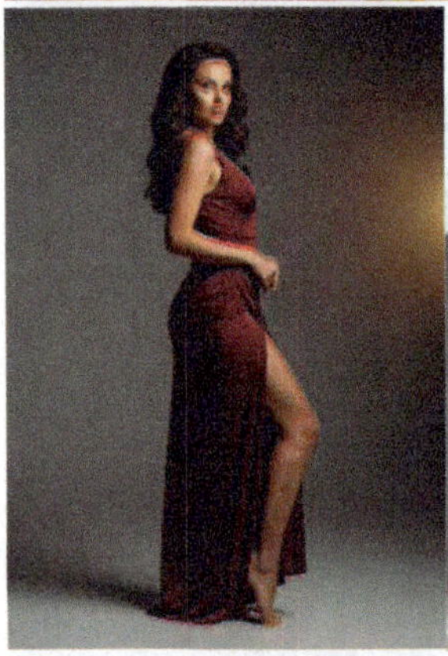
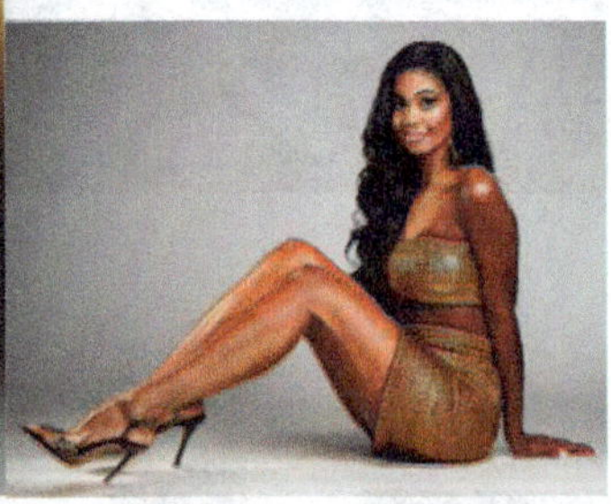
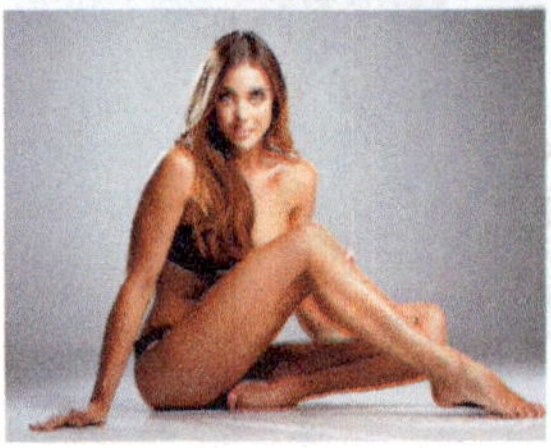

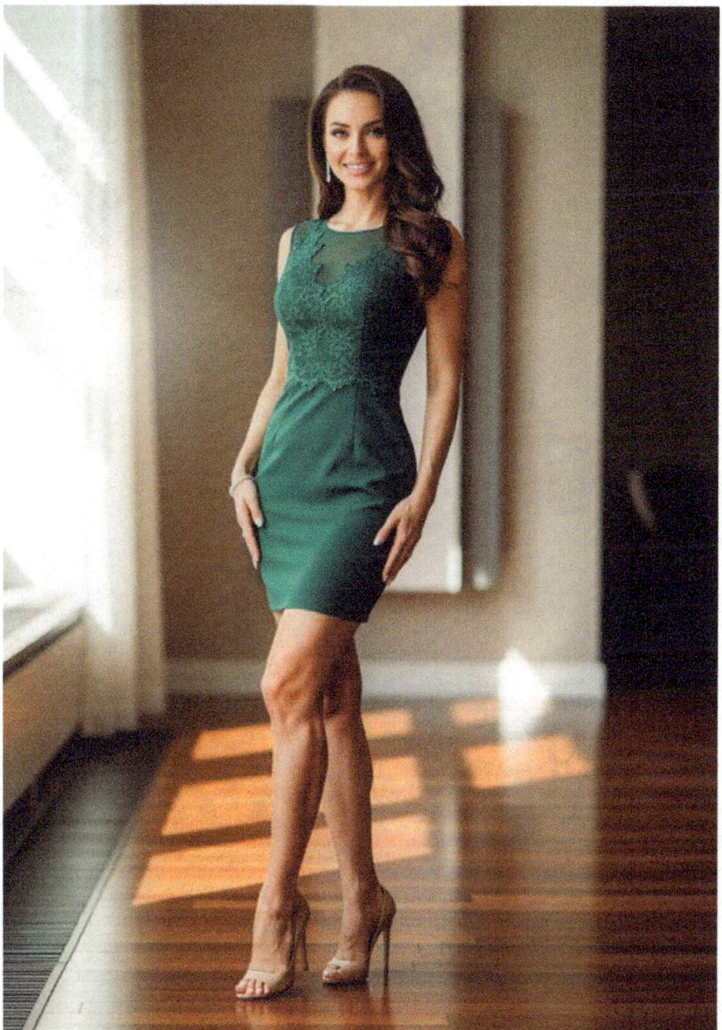

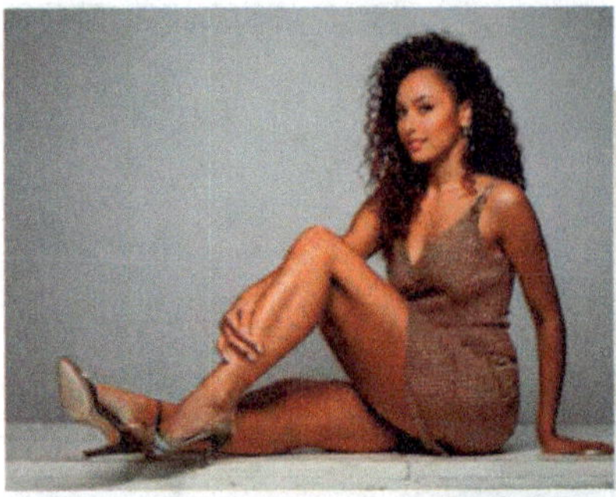
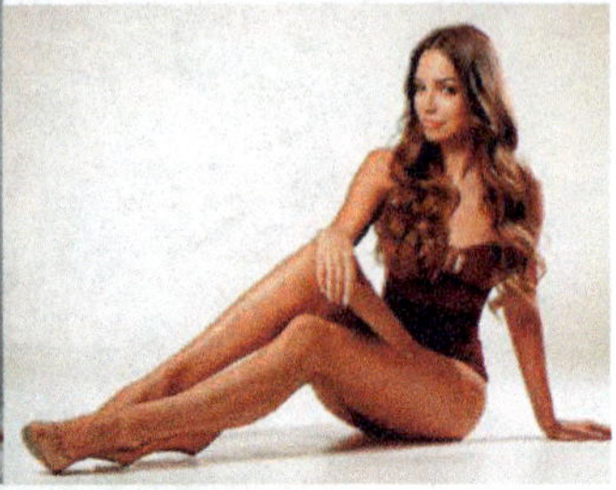
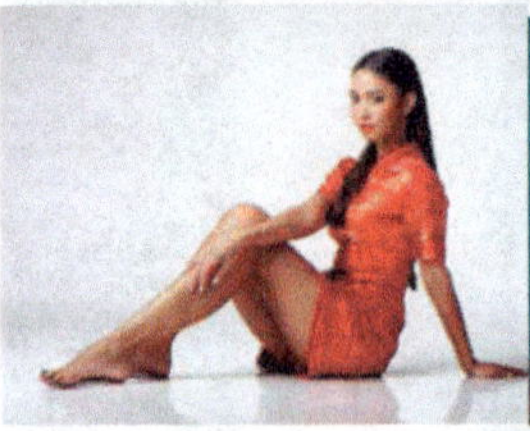
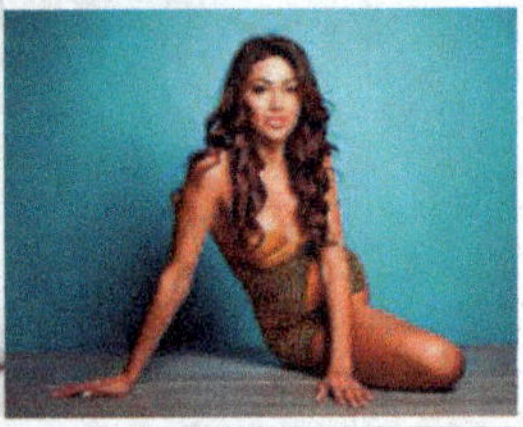
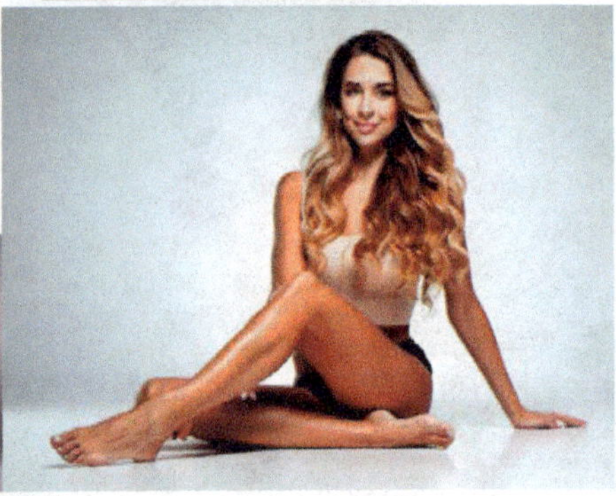
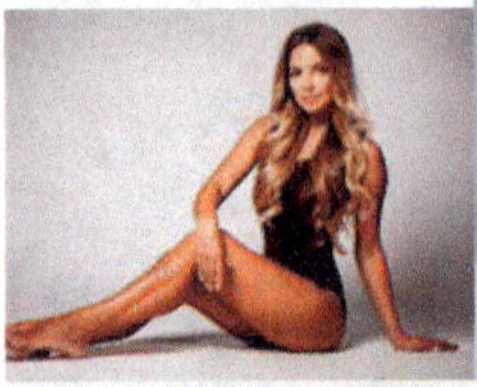
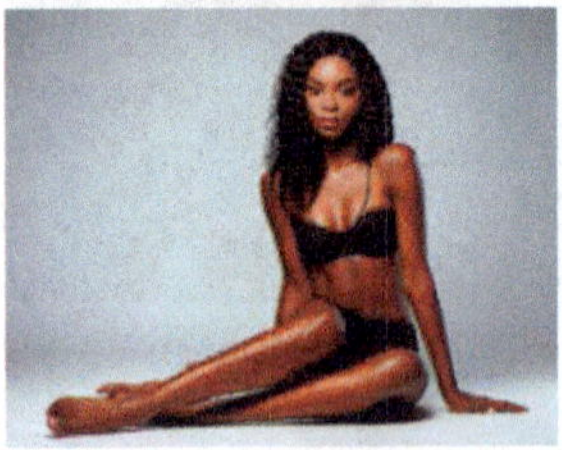

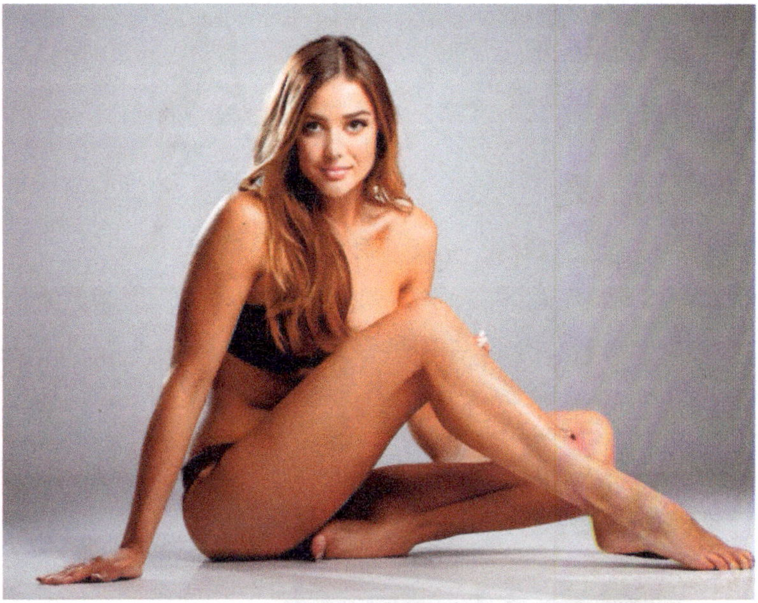

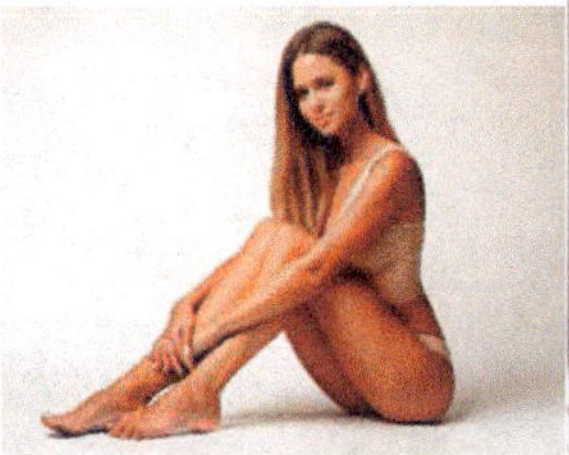

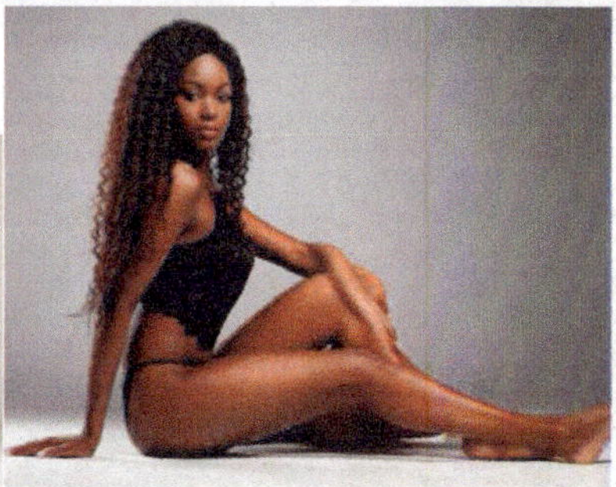

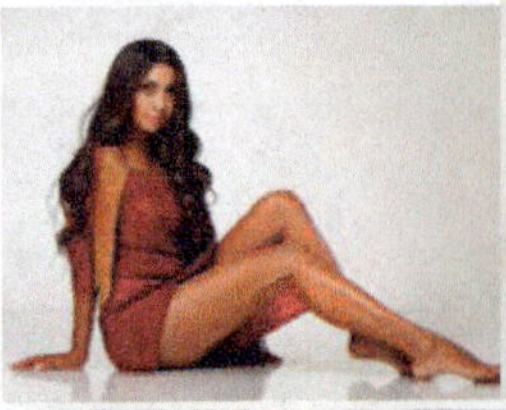
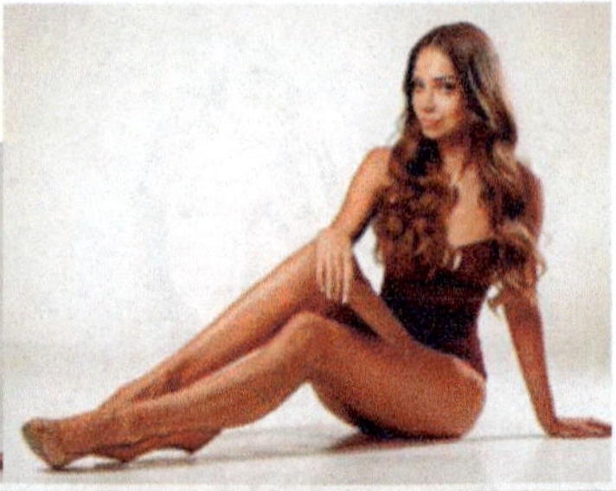
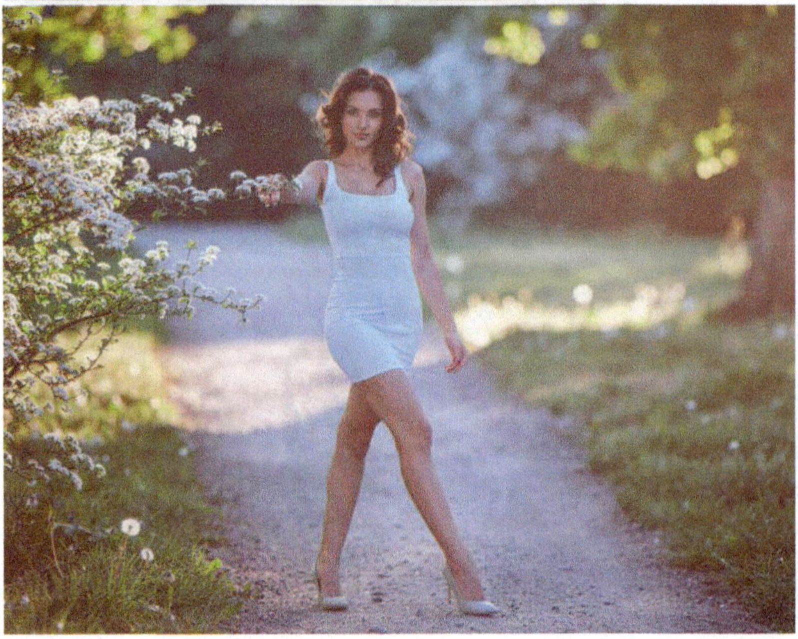

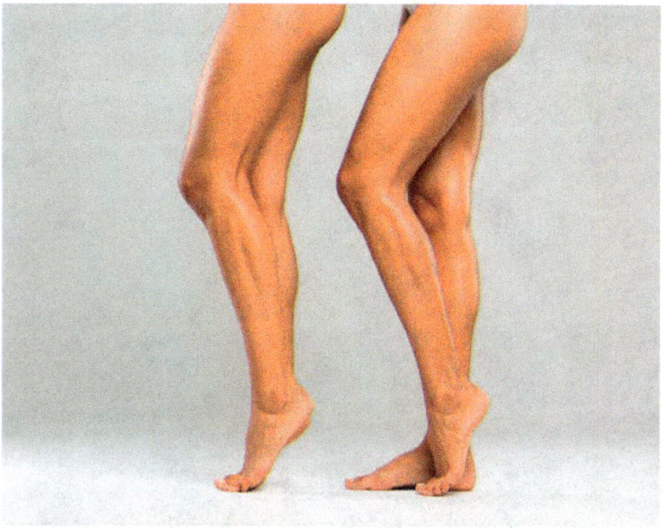

Legs have made many images in the beauty business as parts that were well shaped or attractive. I wonder how much my concepts of beauty were tainted by seeing these images. I think about the differences now that stockings and pantyhose aren't in favor as daily wear.

THE FOLLOWING PAGES ARE PAGES FROM THE BOOK "THE FEMALE FORM FROZEN" AND ARE COLLAGES THAT ARE ONE OF A KIND IN ORIGINAL PAPER AND COLOR FOR SALE FOR 1500 DOLLARS A PIECE

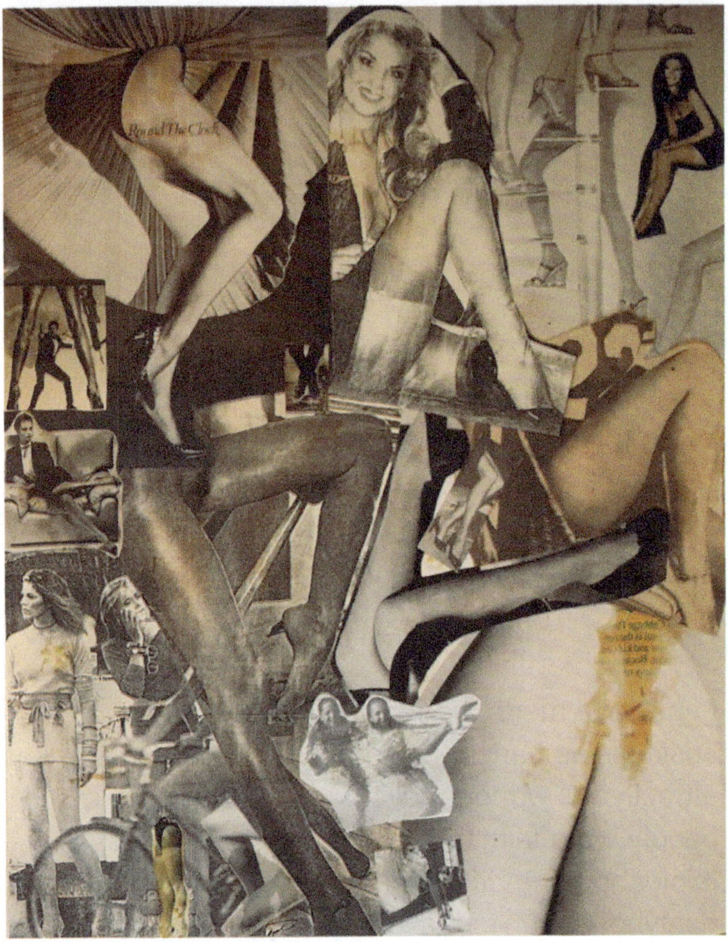
Round The Clock

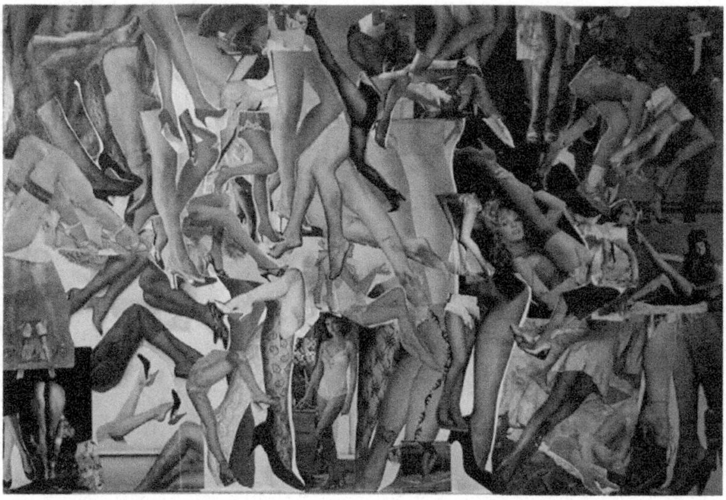

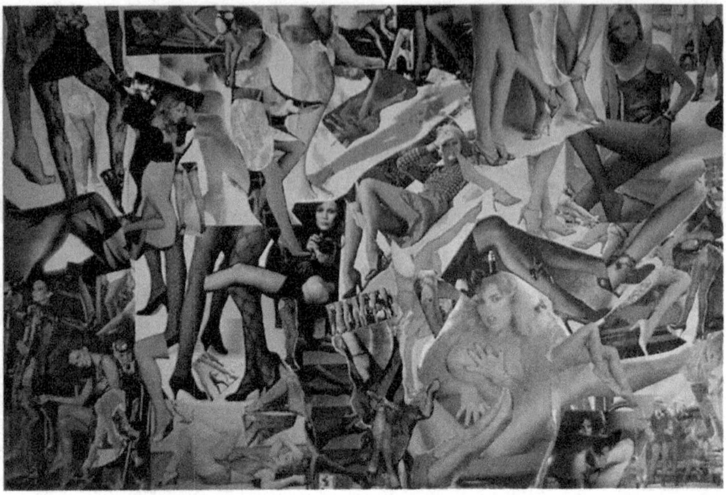

LEGS FROM STOCK FOOTAGE

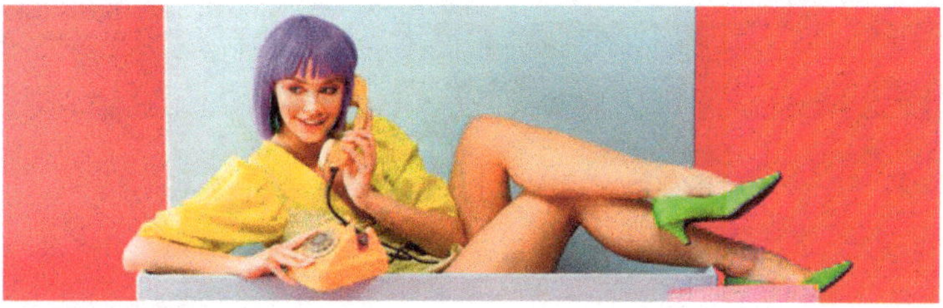

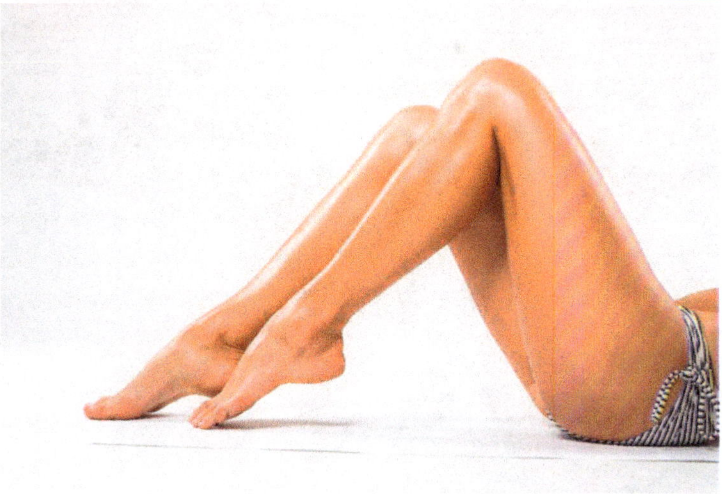

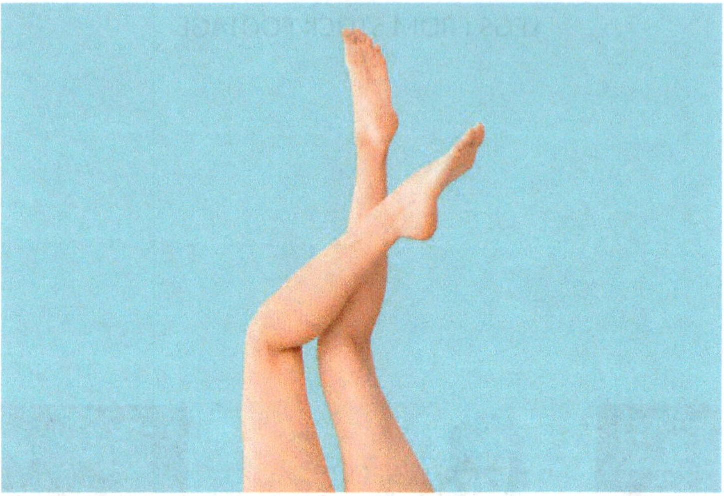

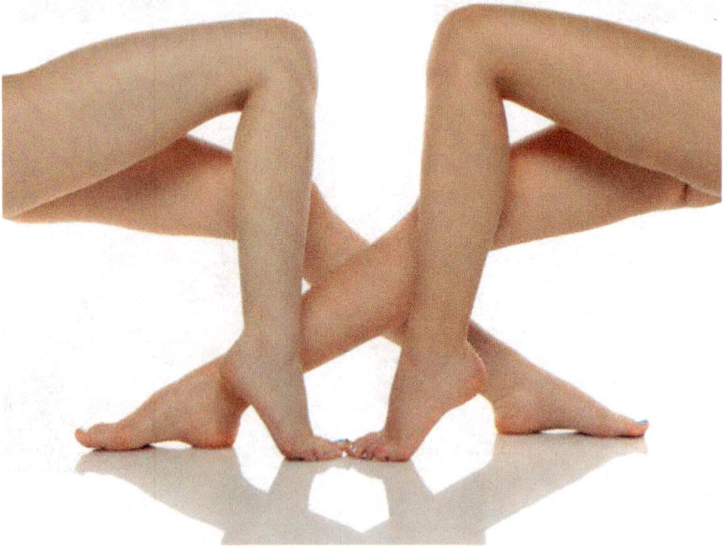

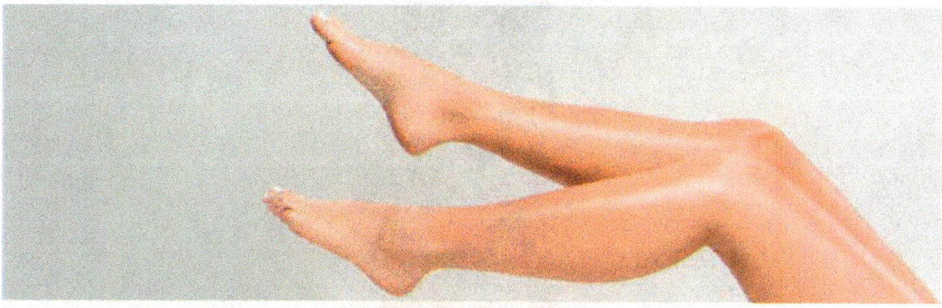

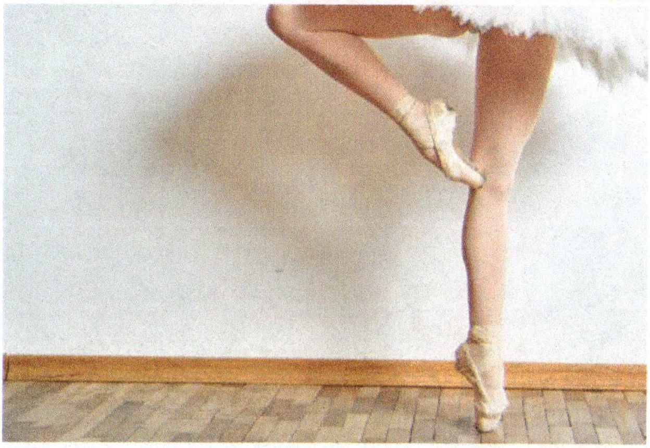

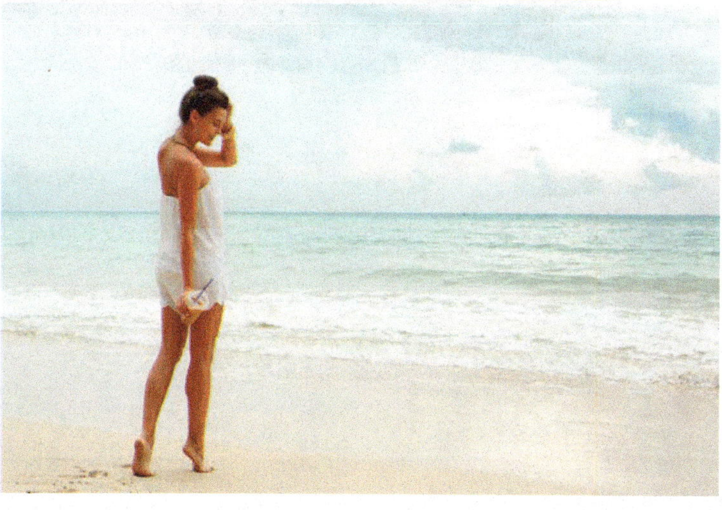

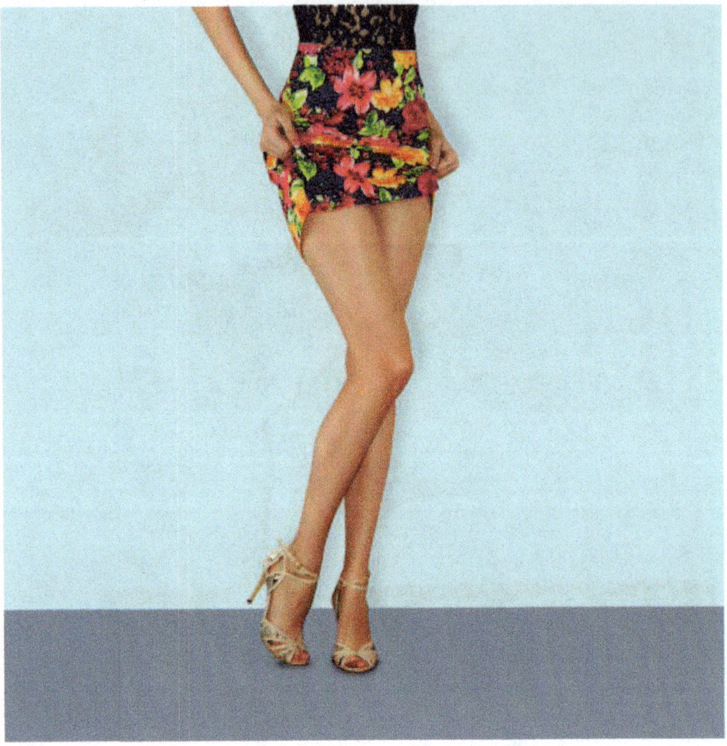

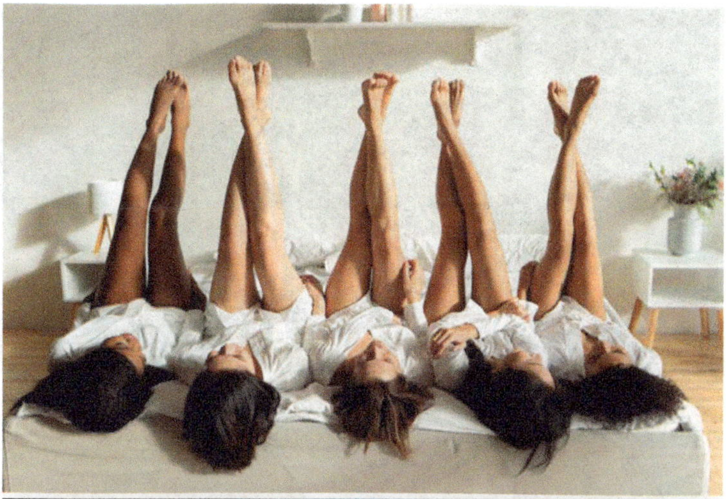

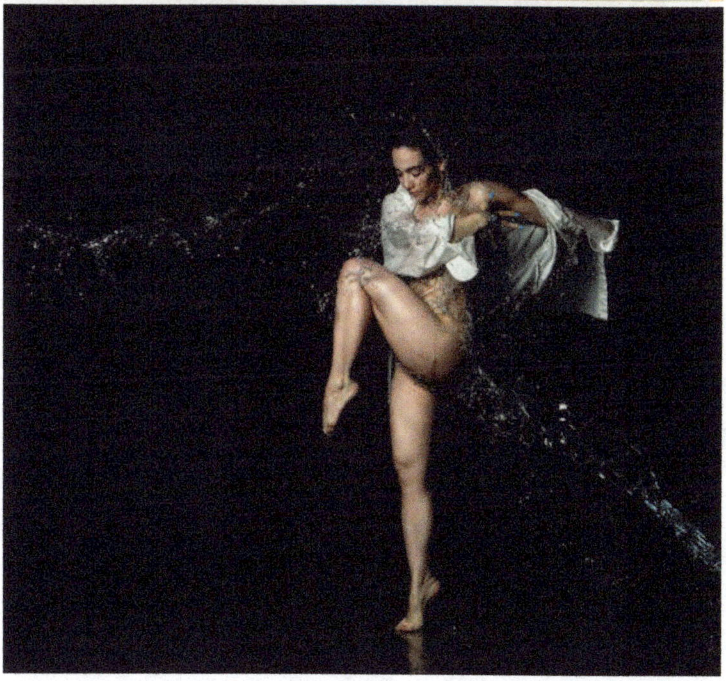

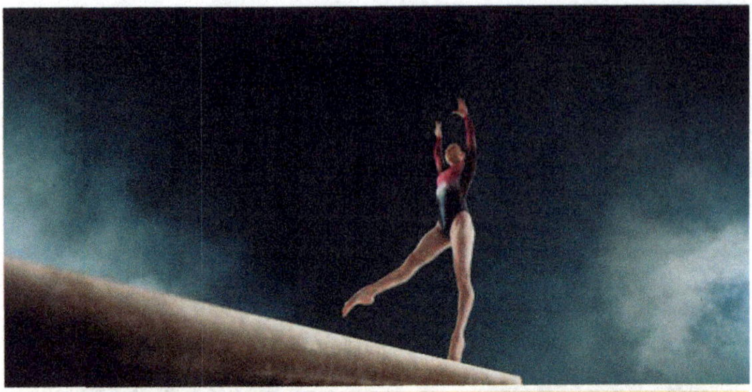

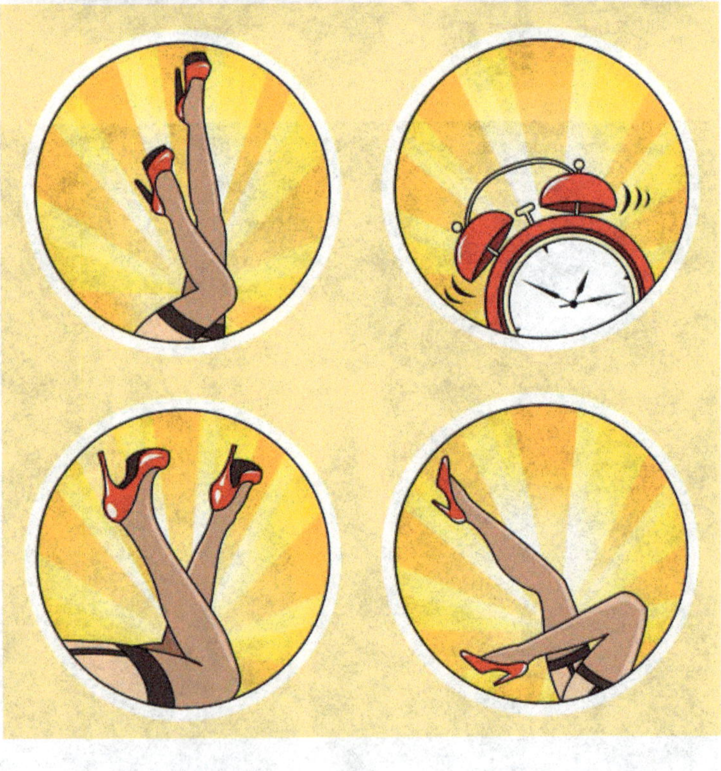

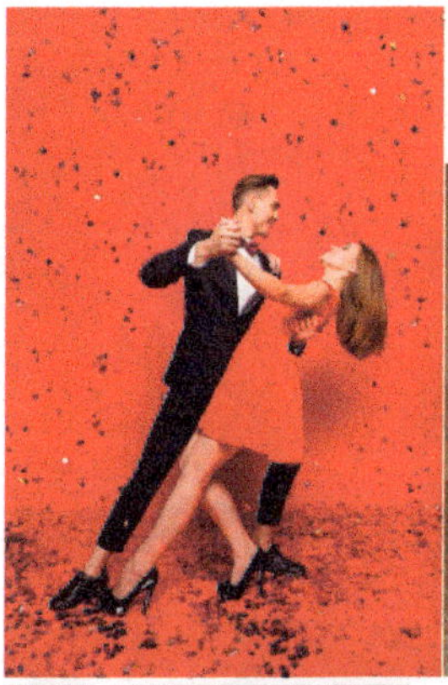

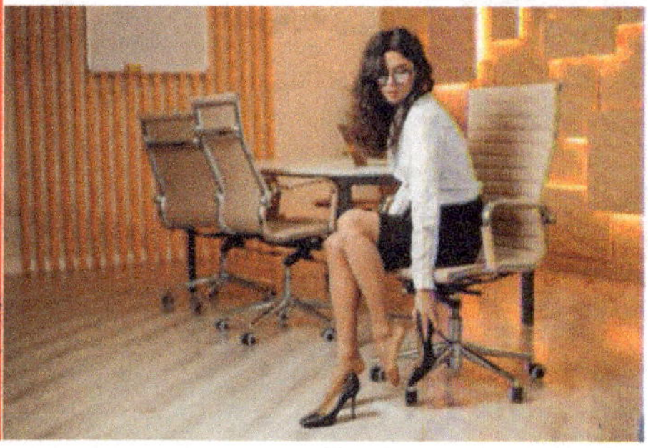

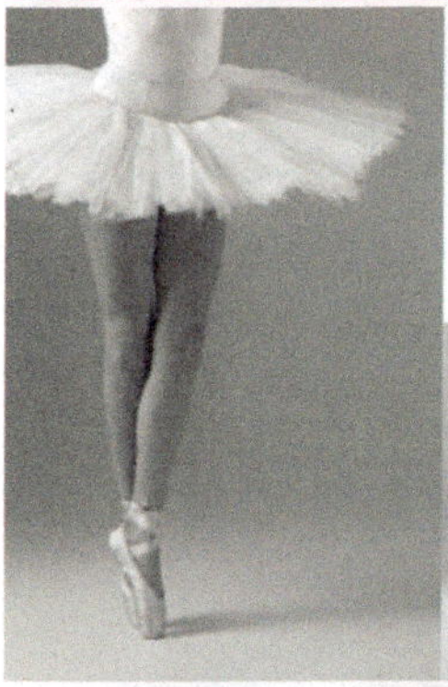

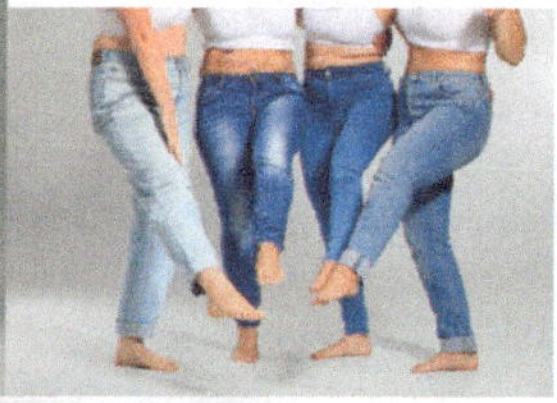

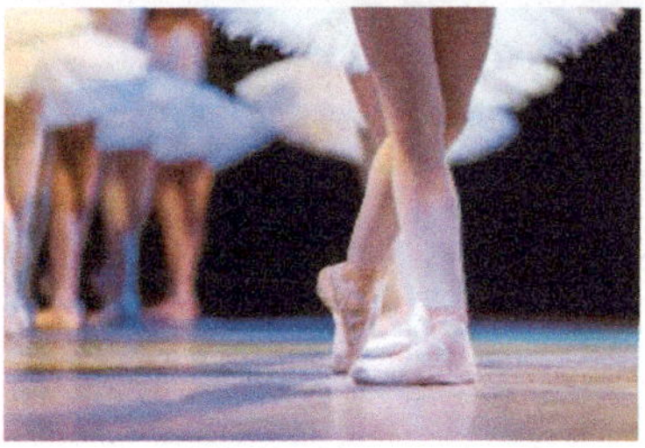

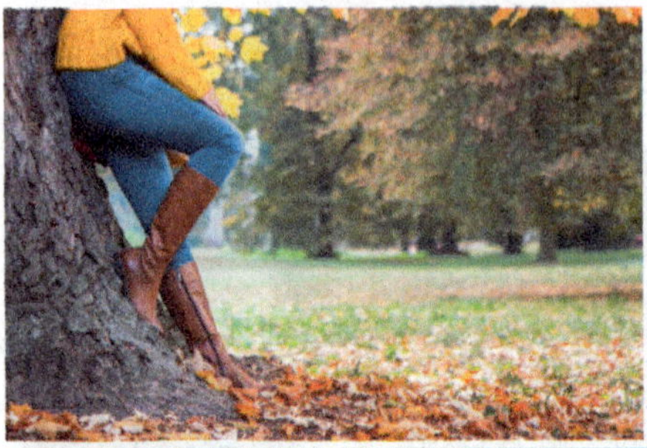

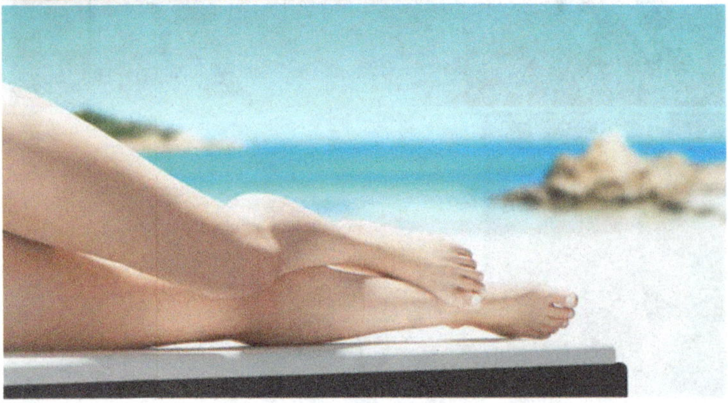

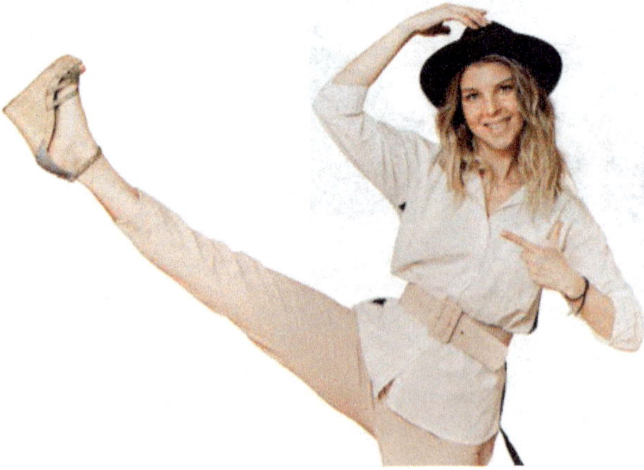

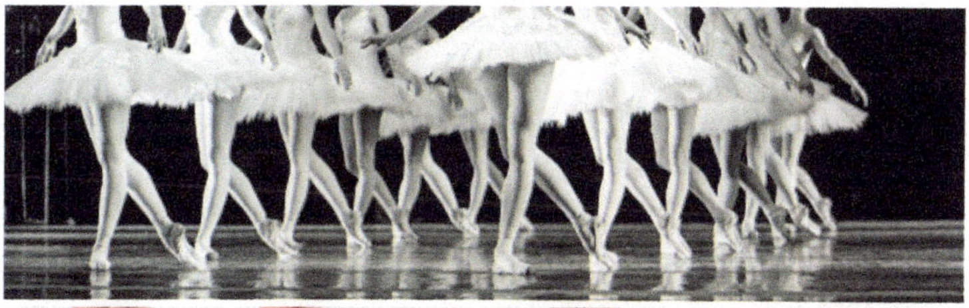

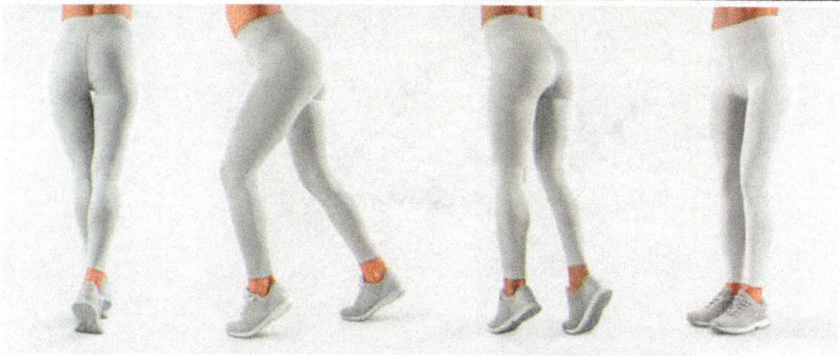

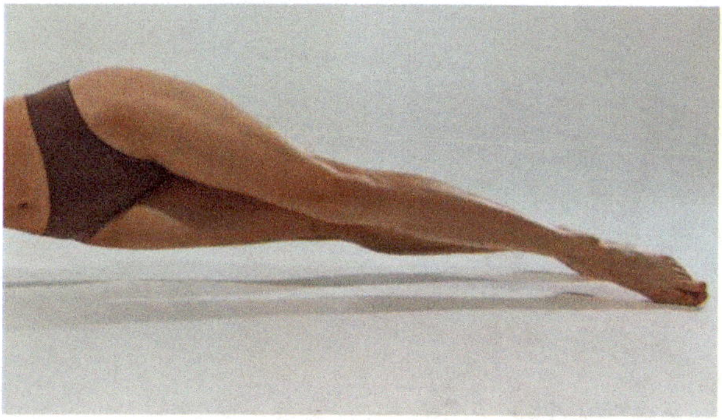

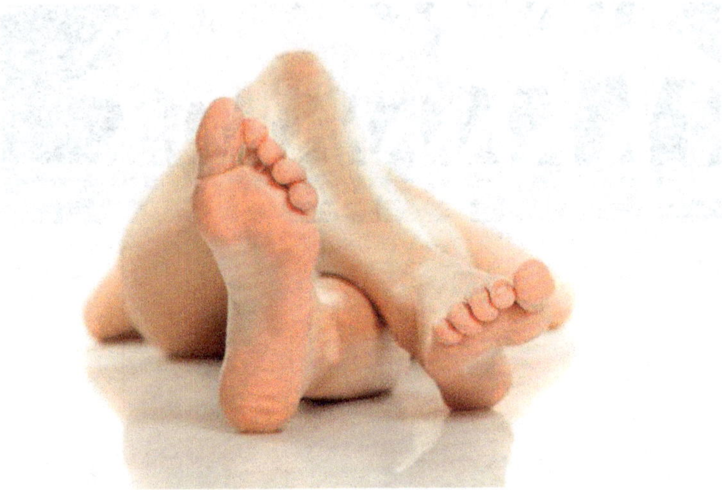

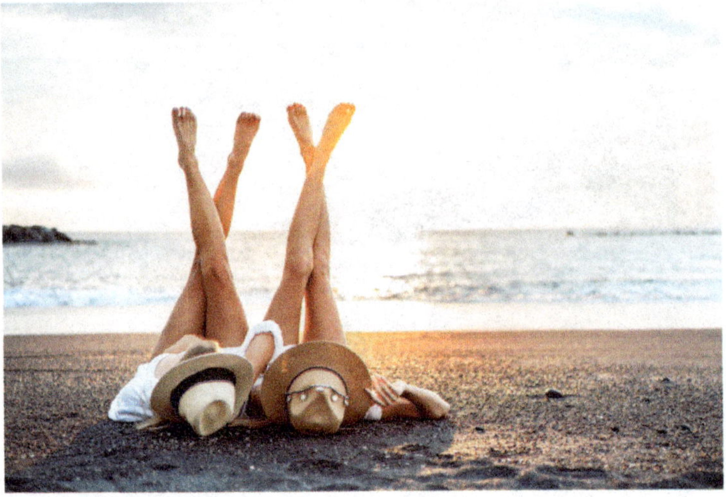

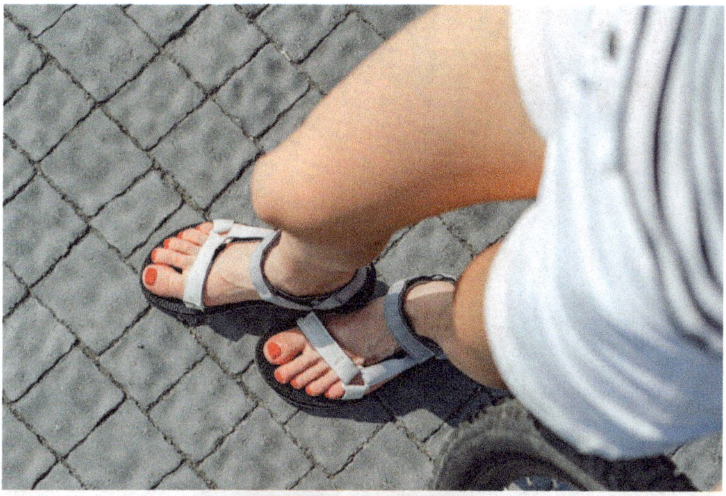

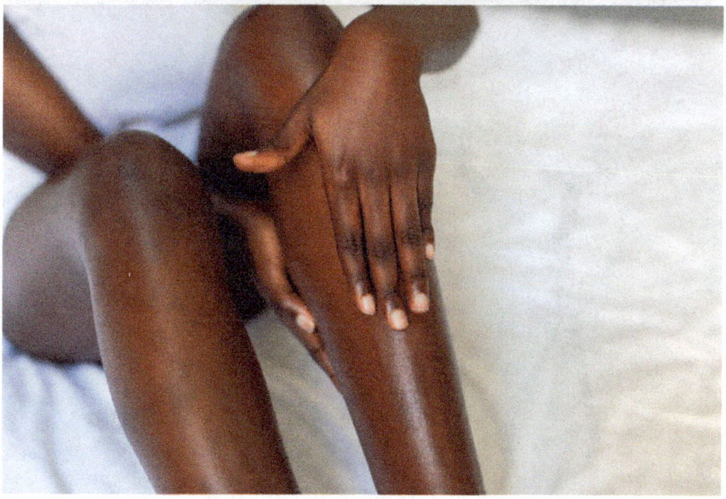

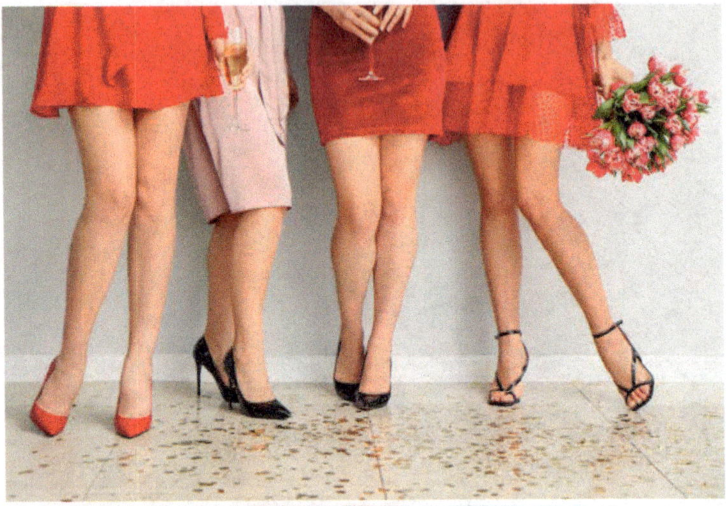

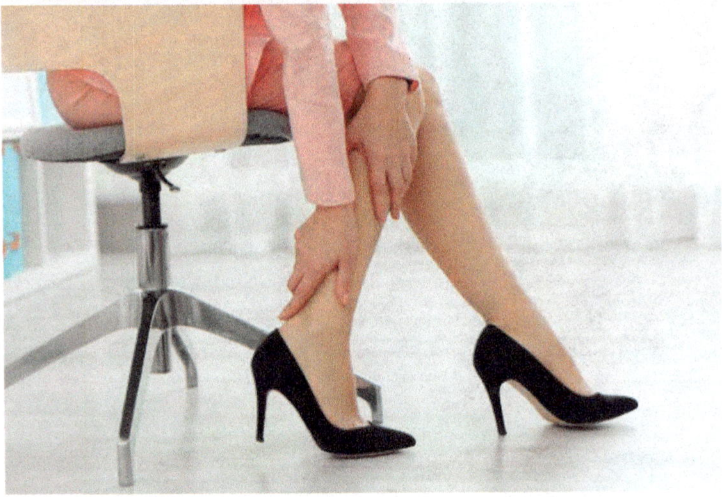

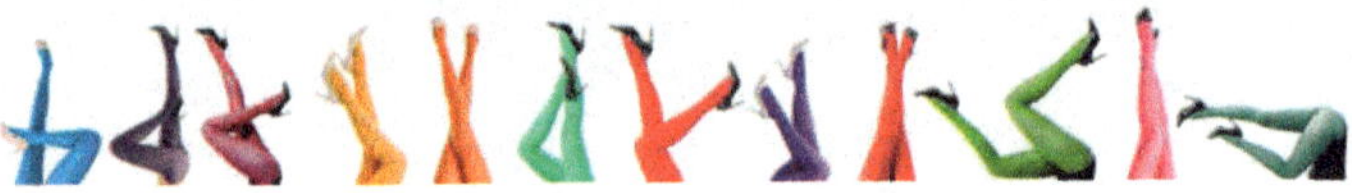

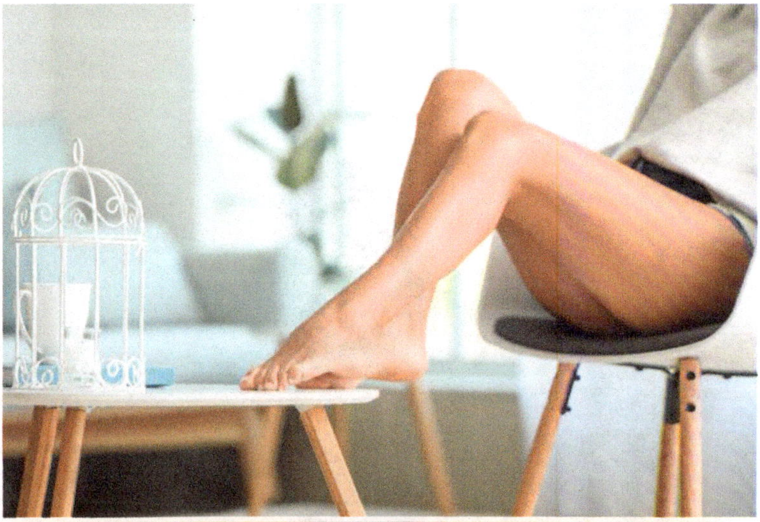

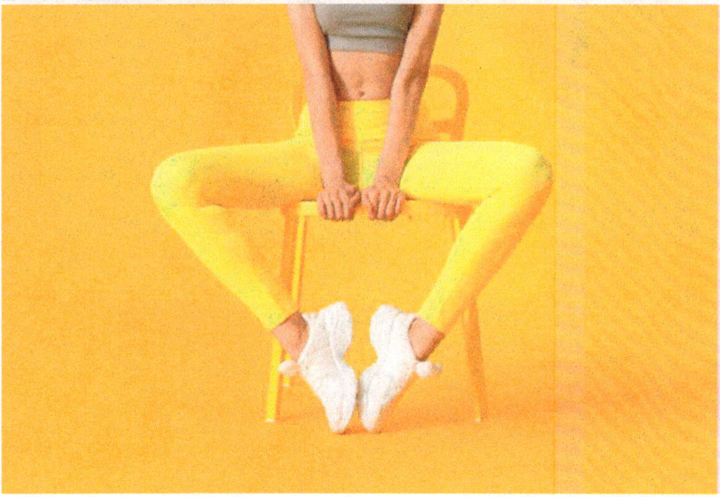

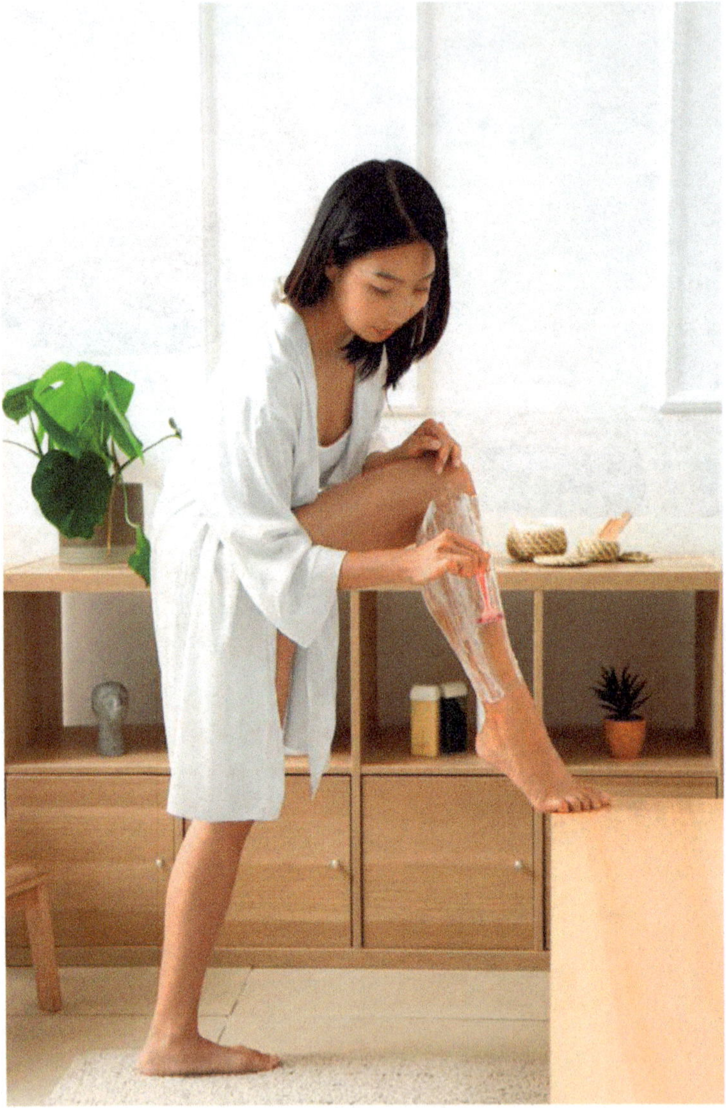

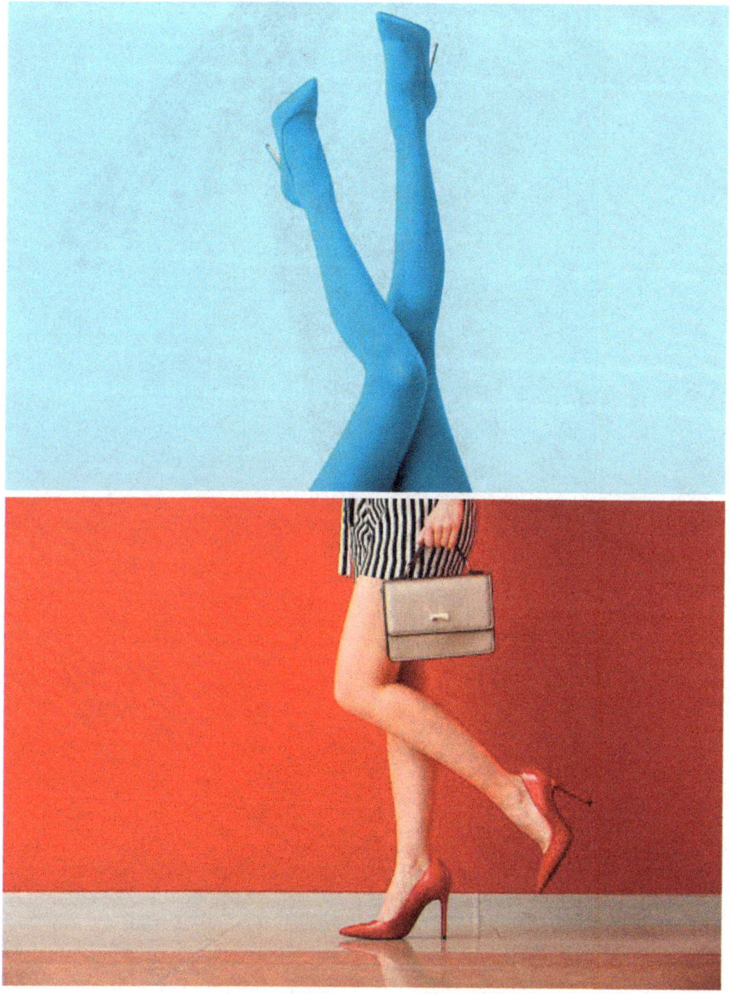

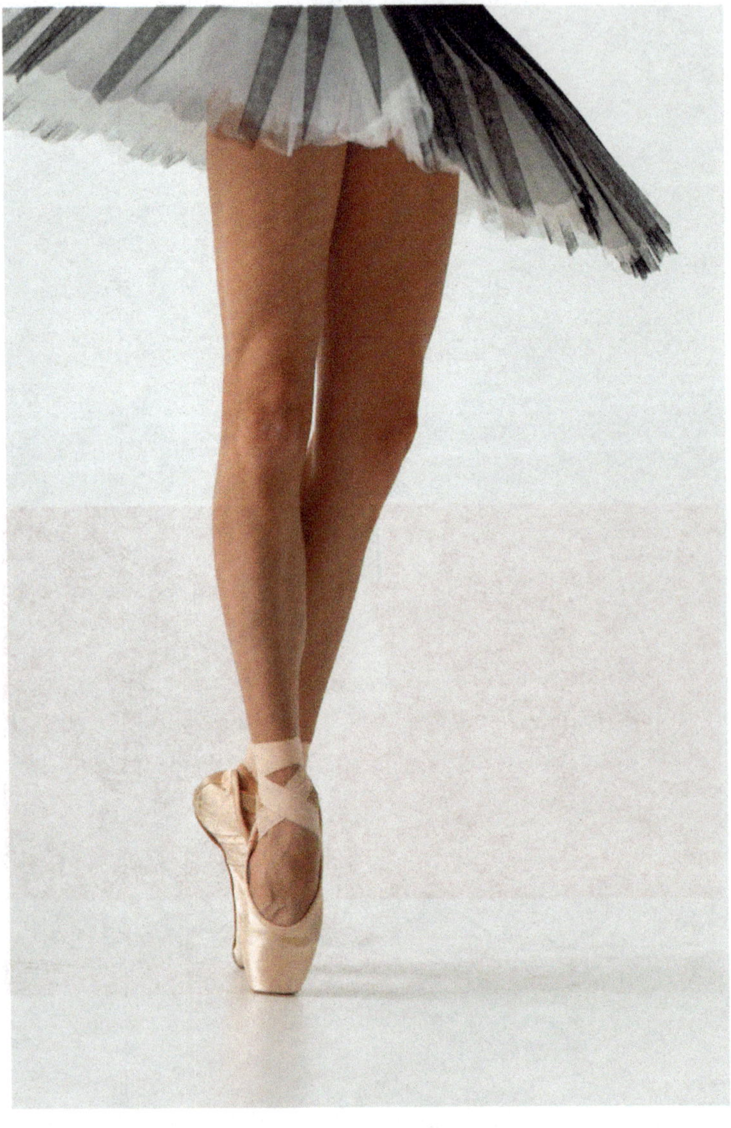

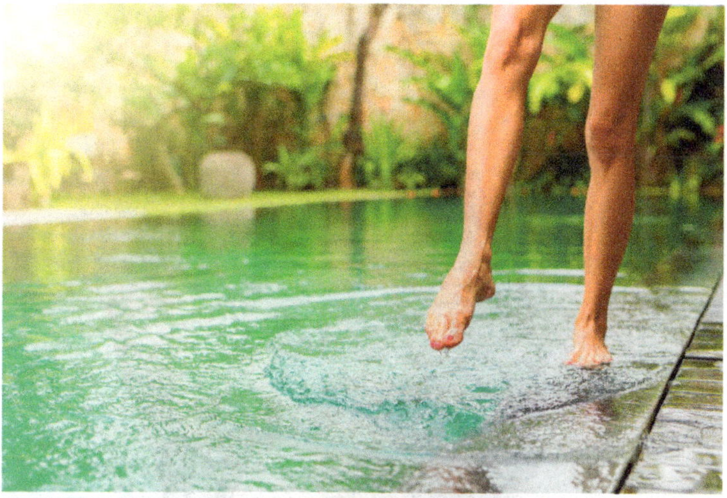

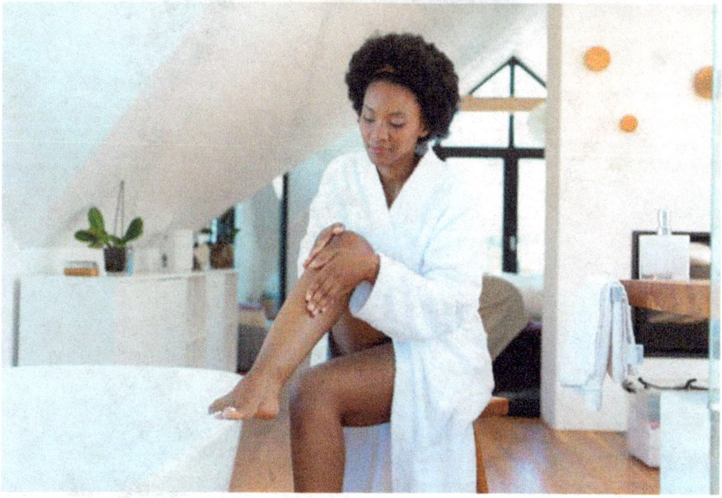

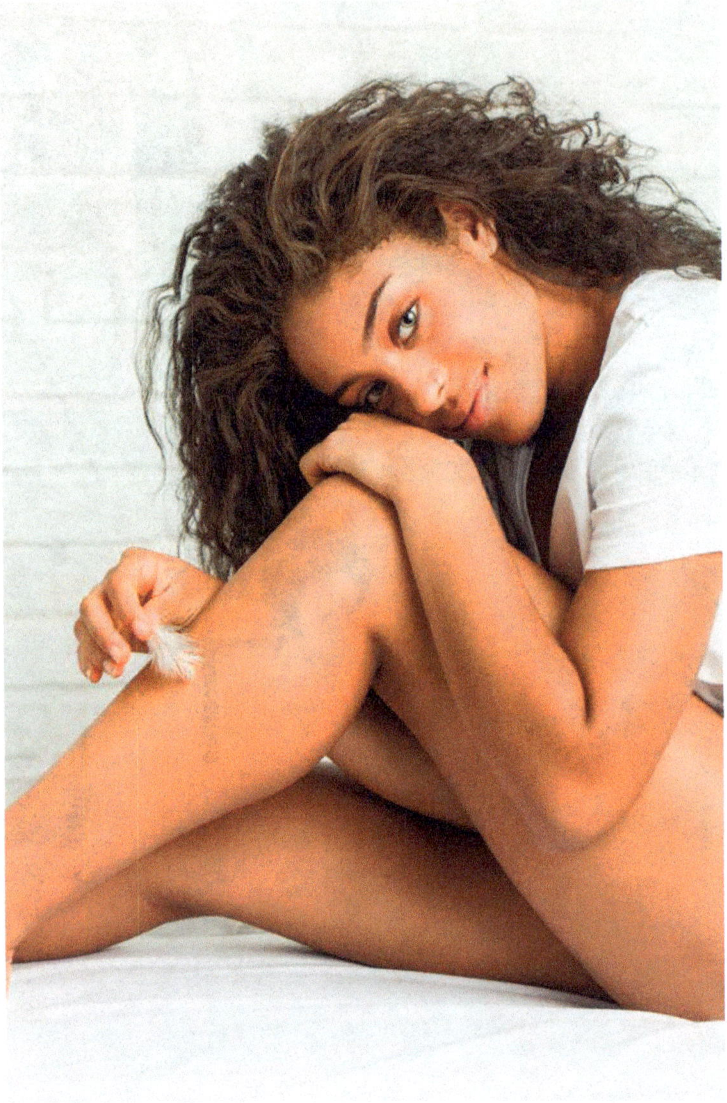

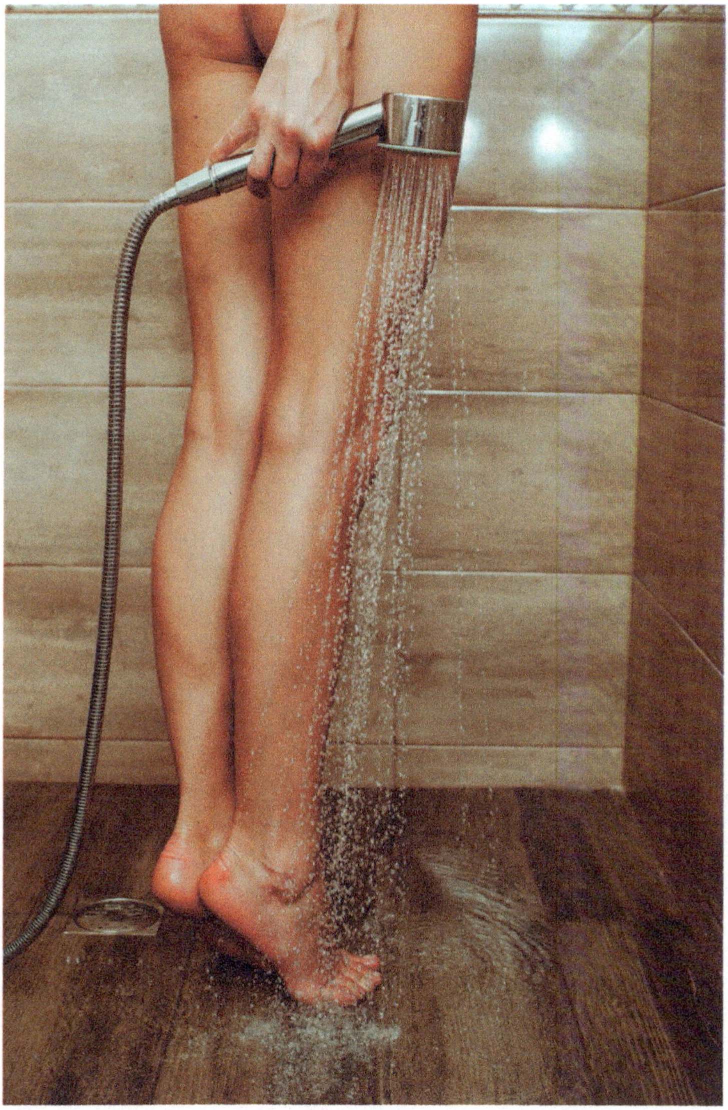

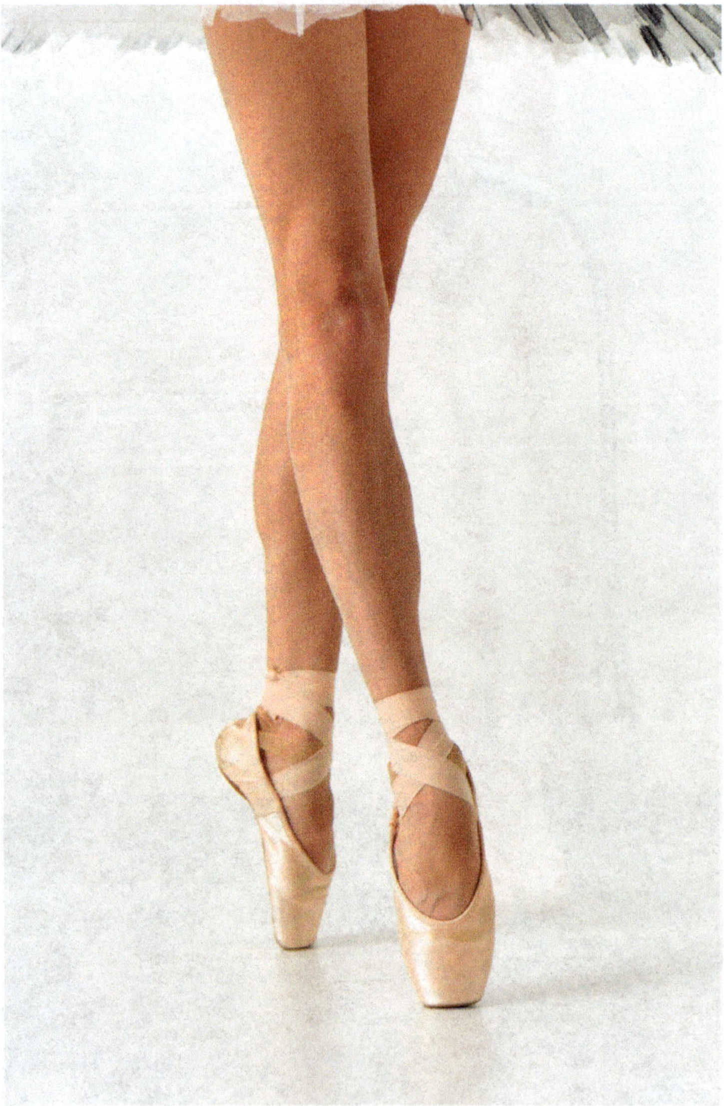

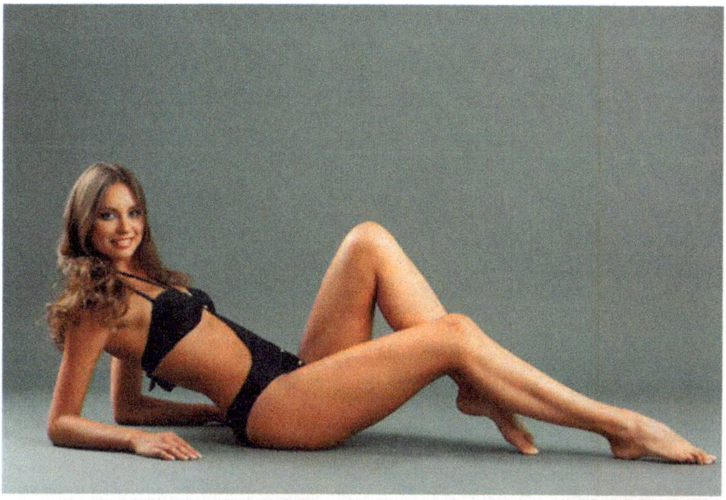

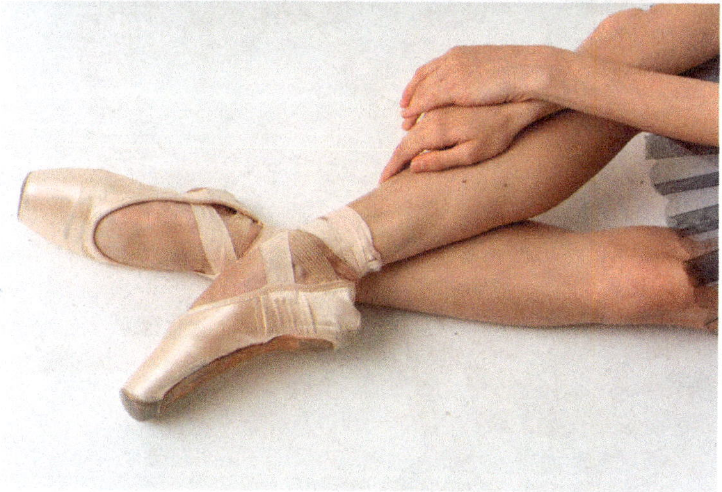

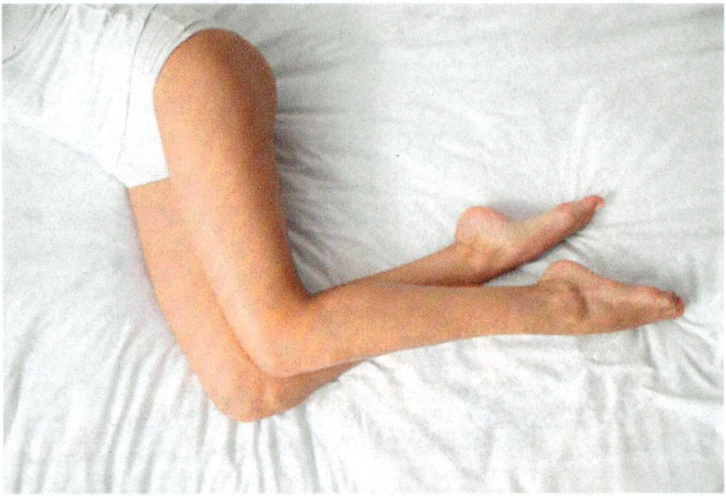

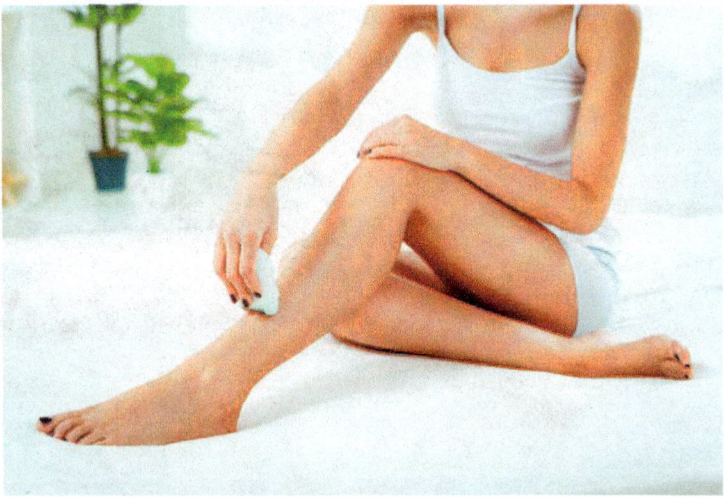

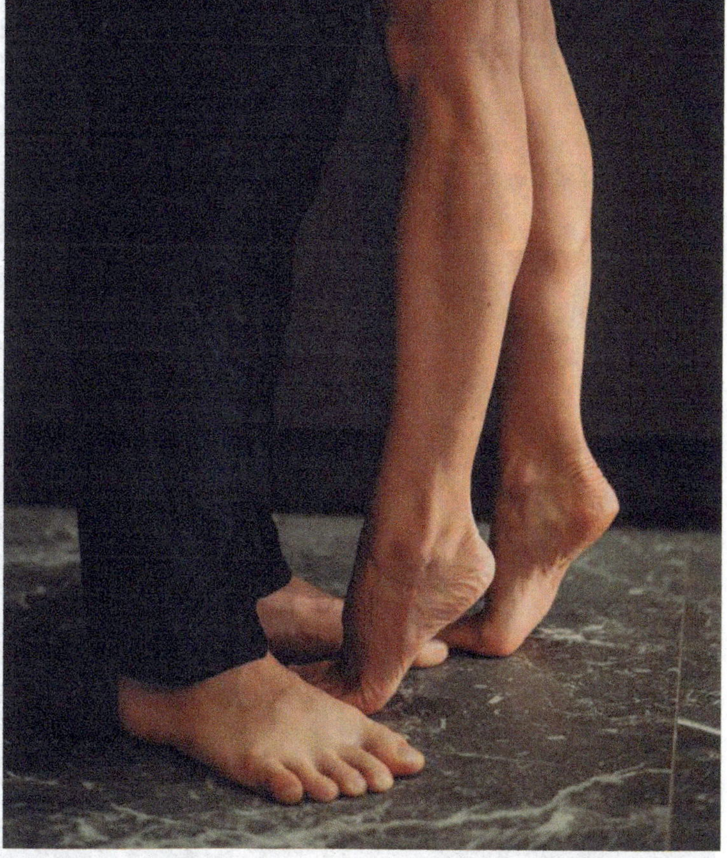

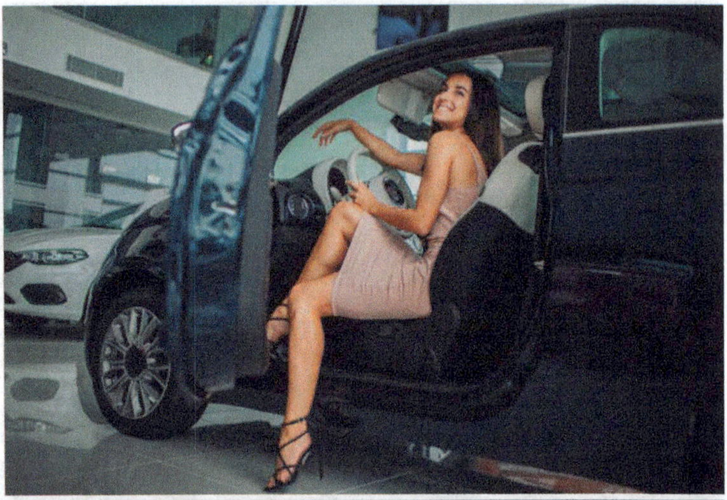

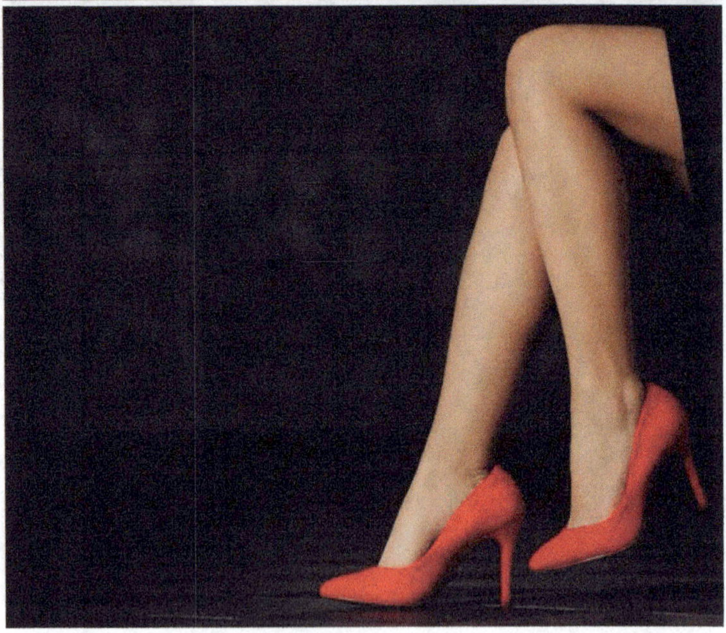

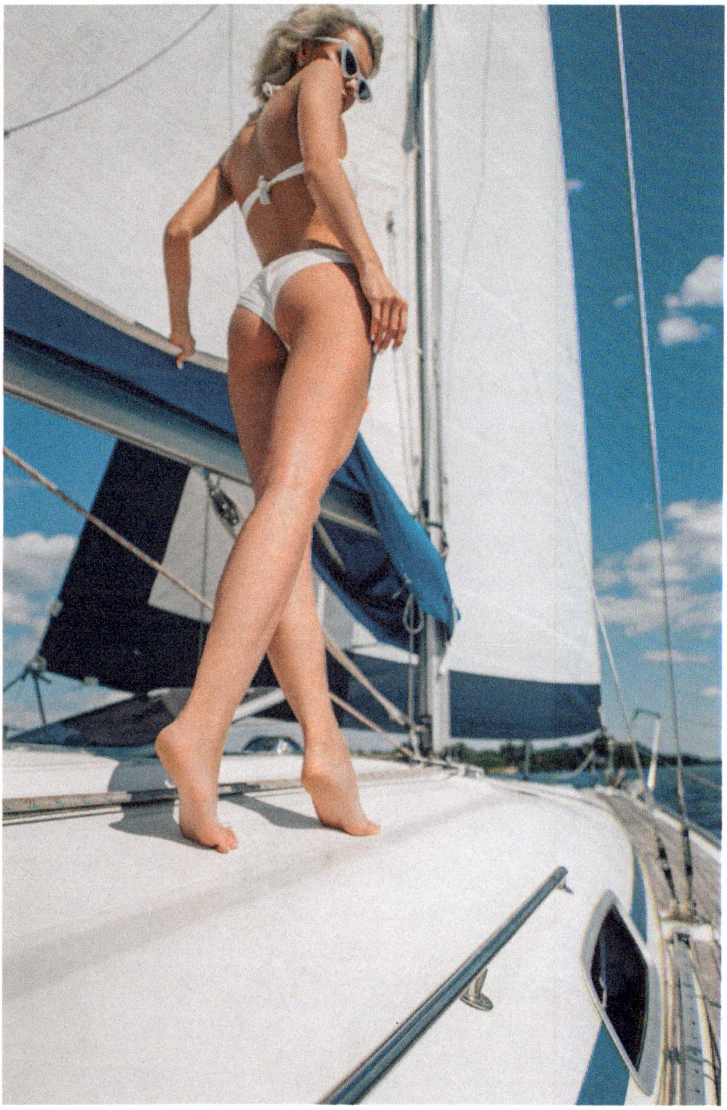

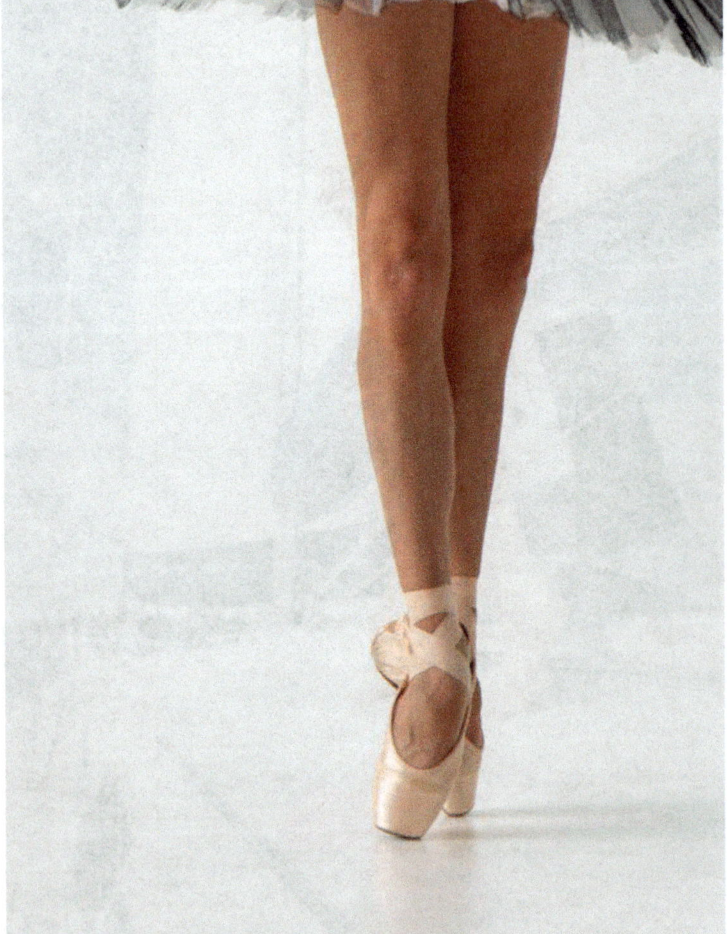

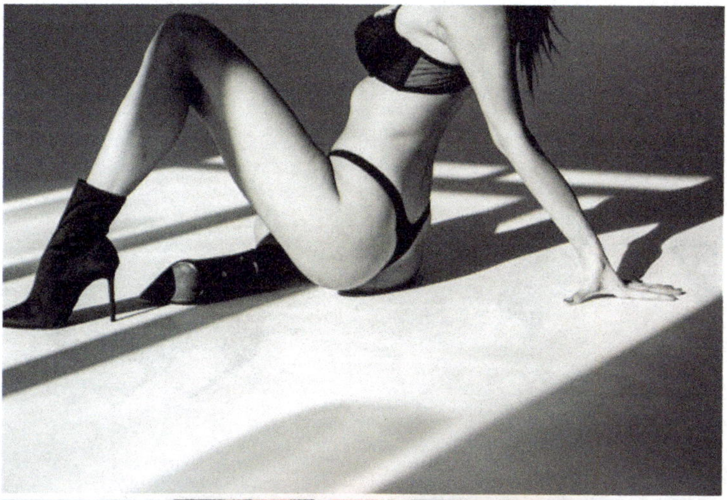
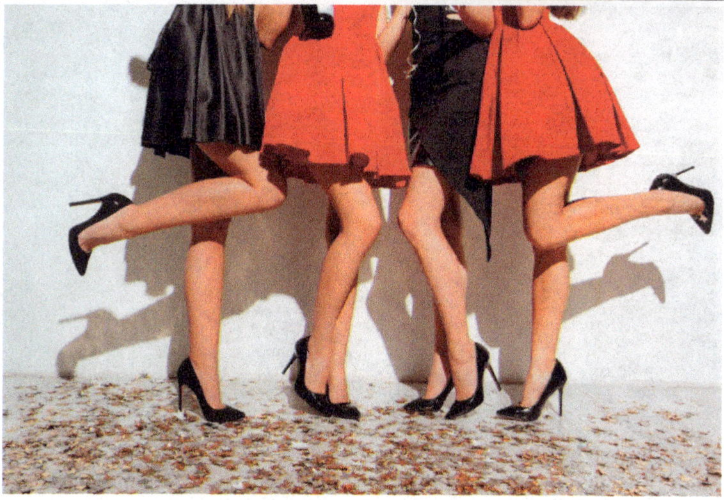

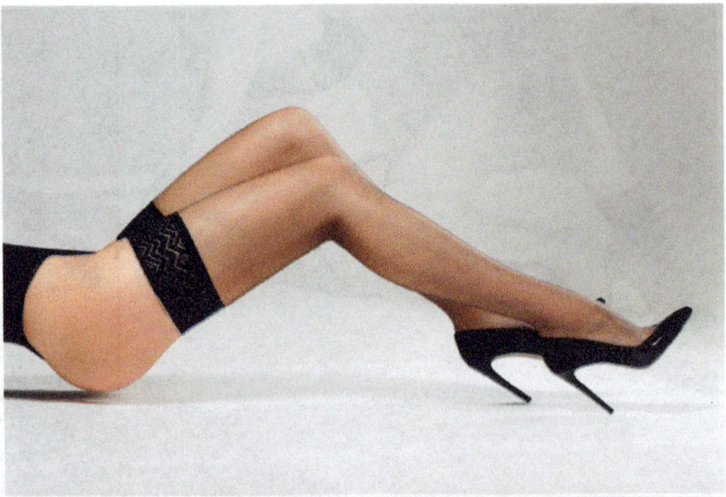

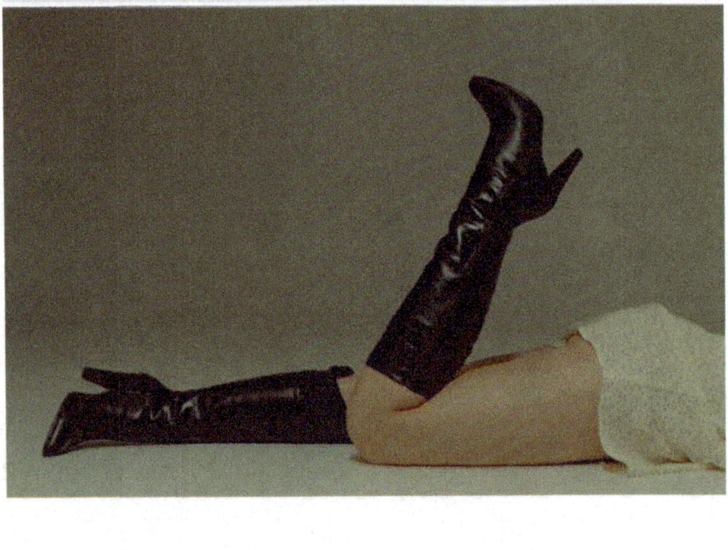

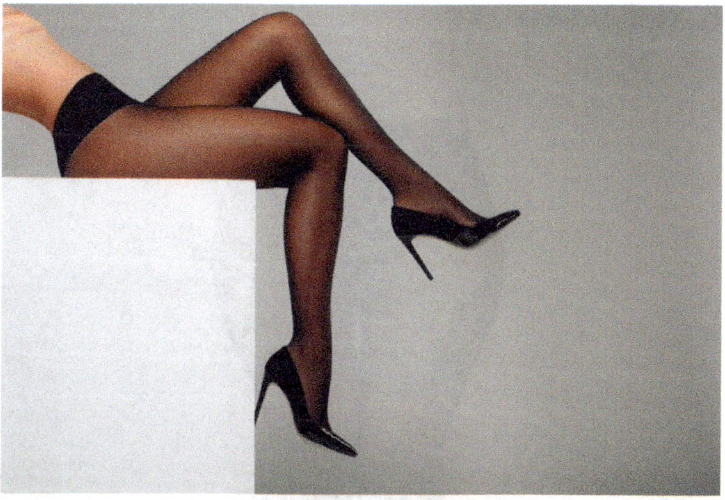

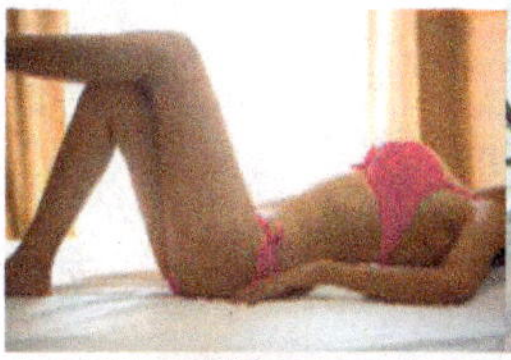

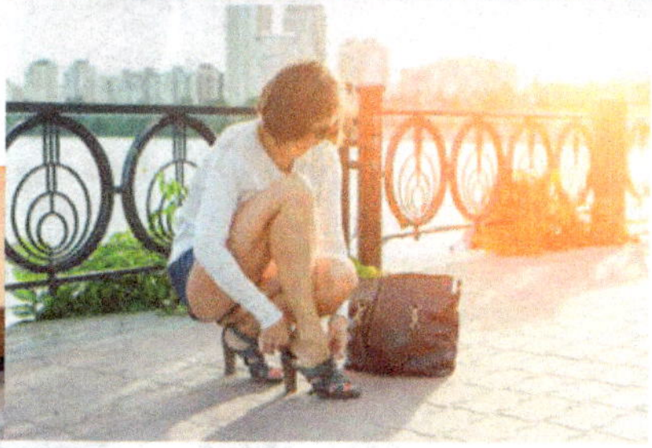

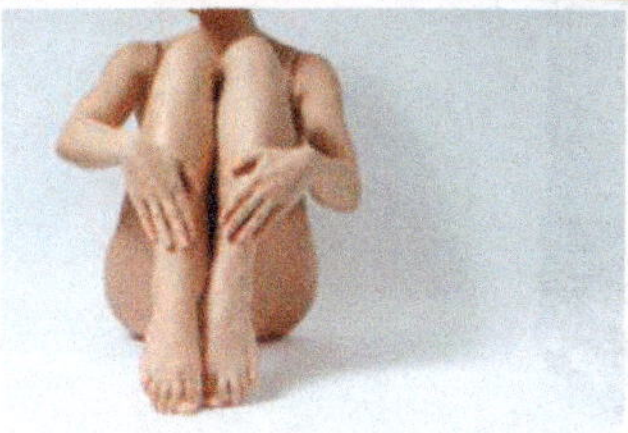

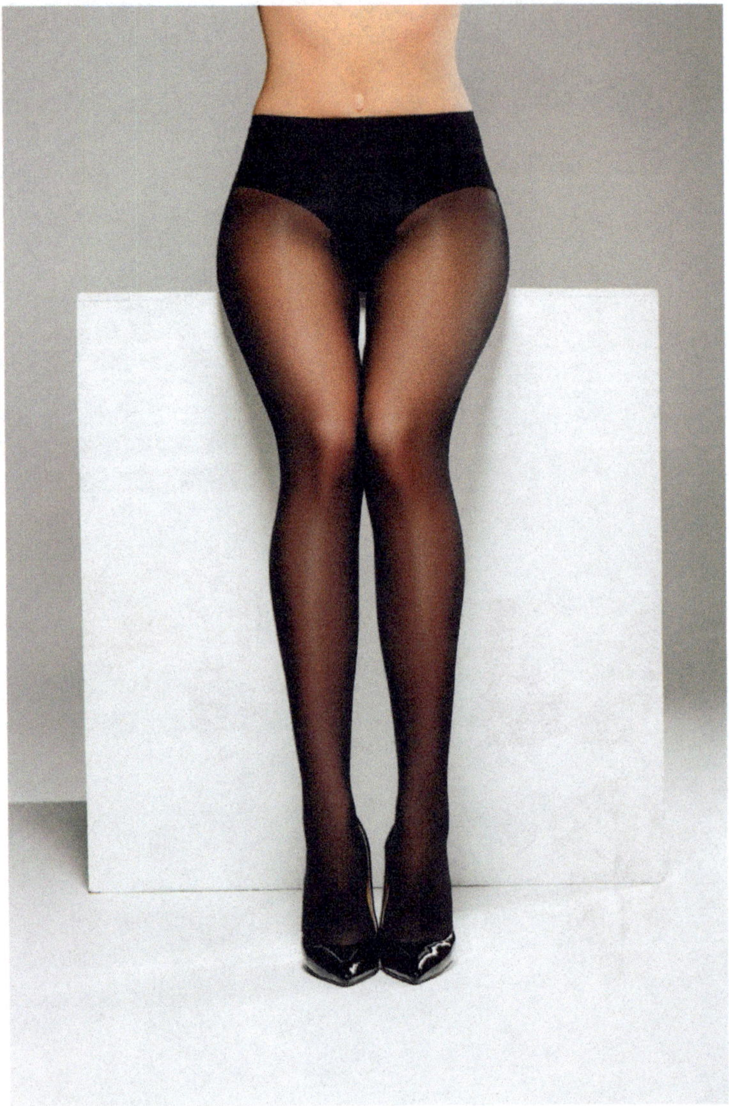

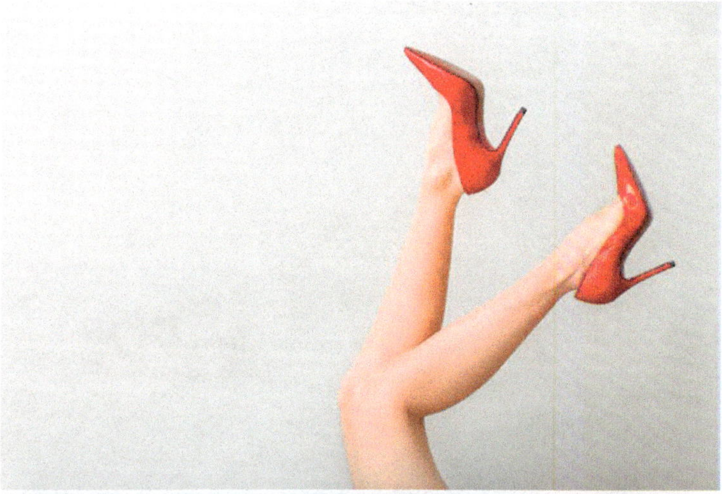
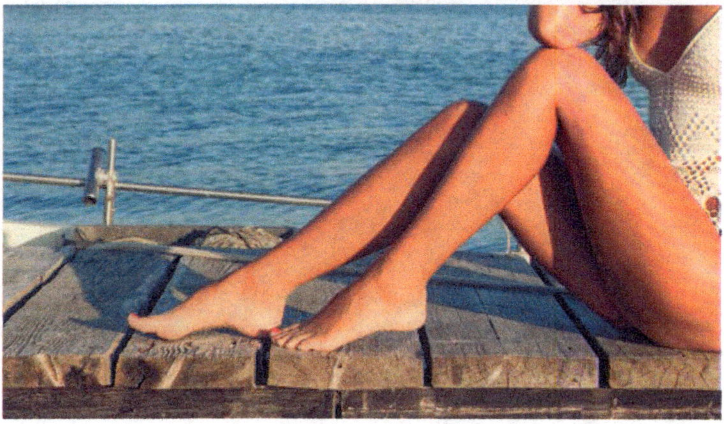

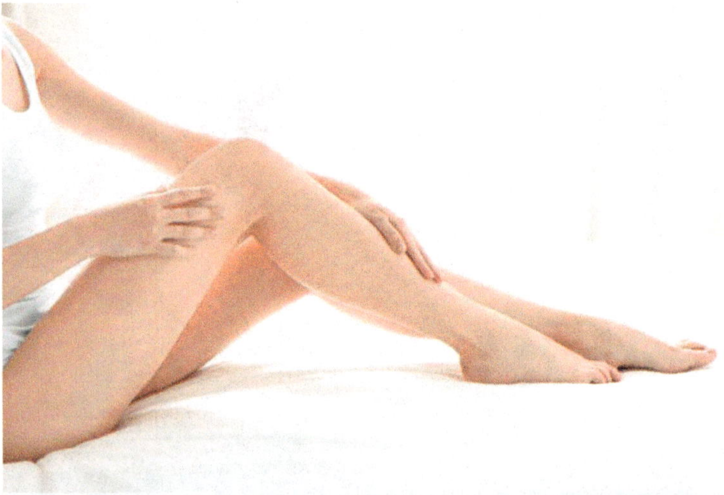

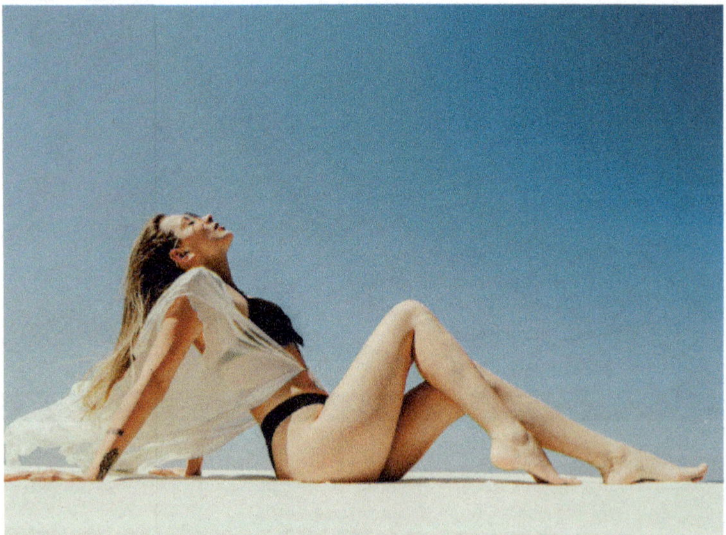

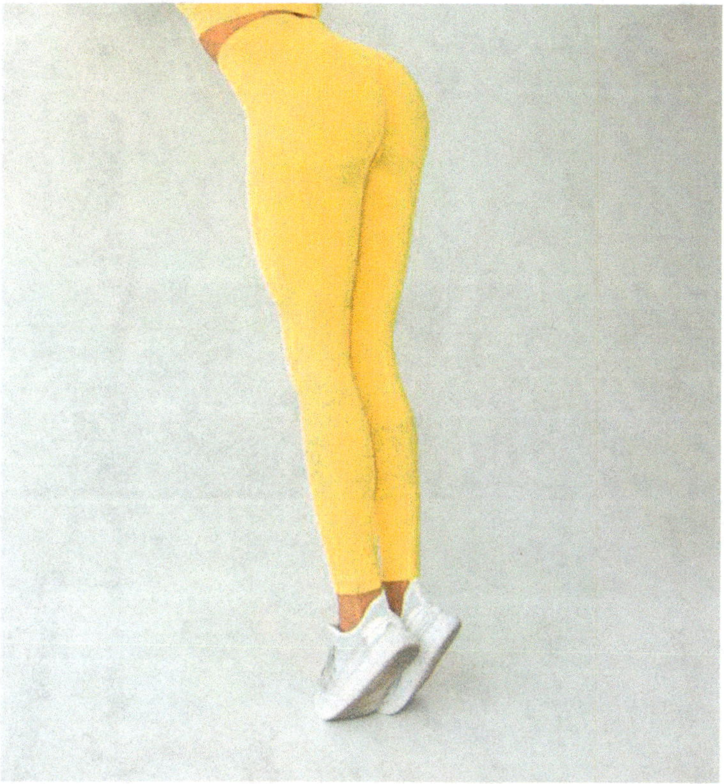

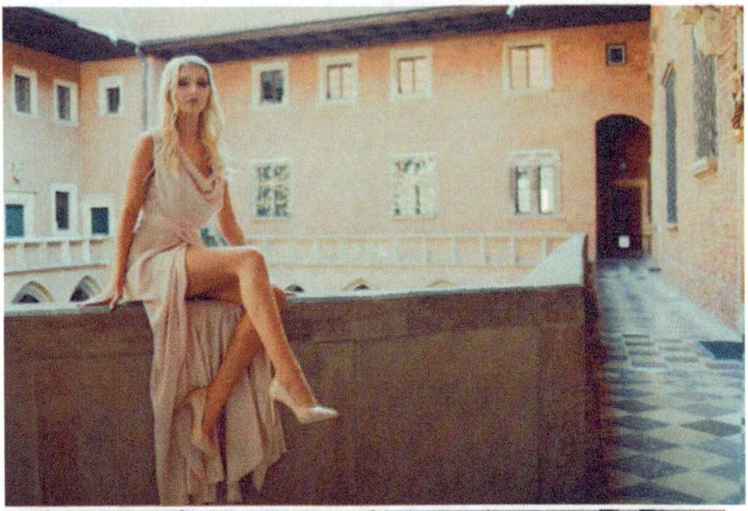

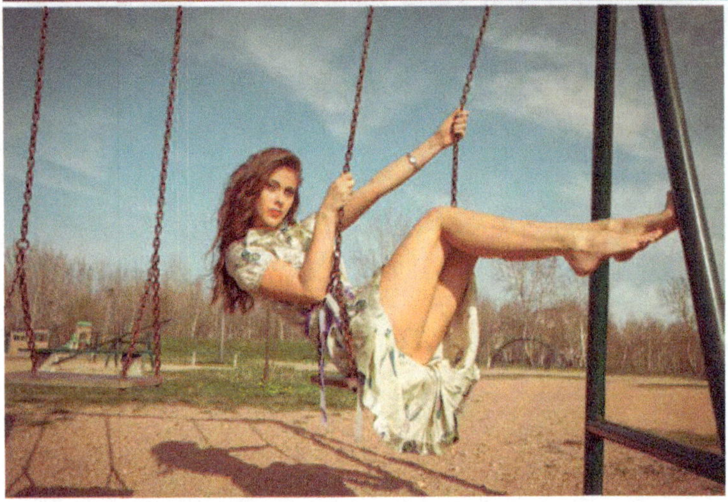

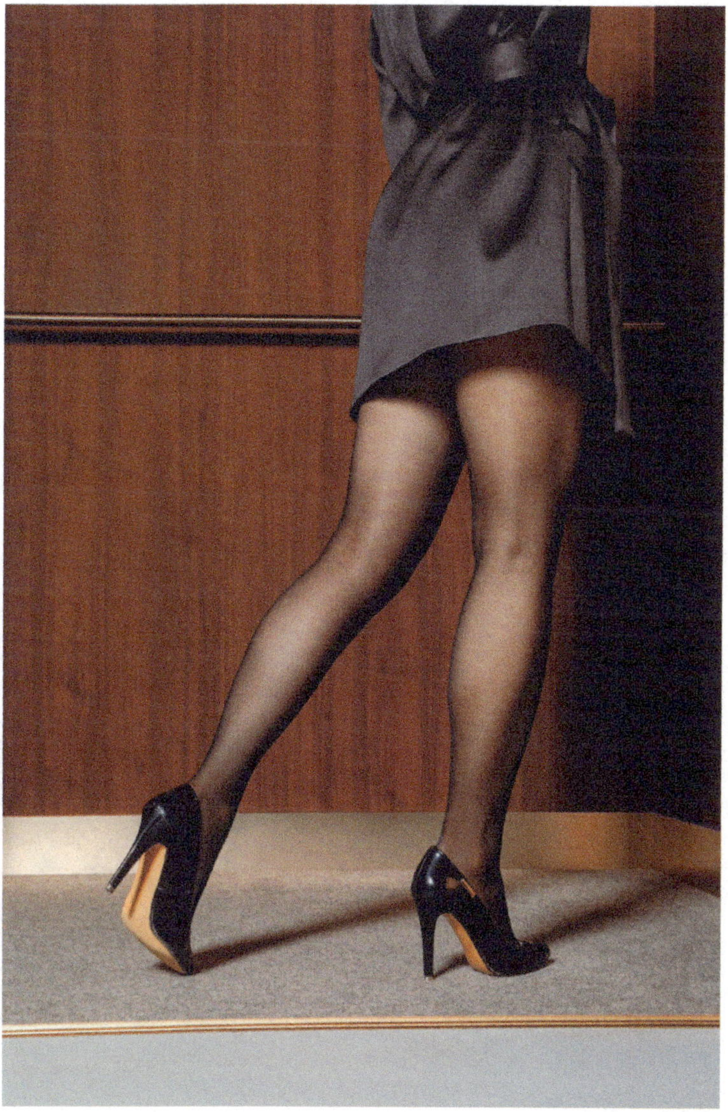

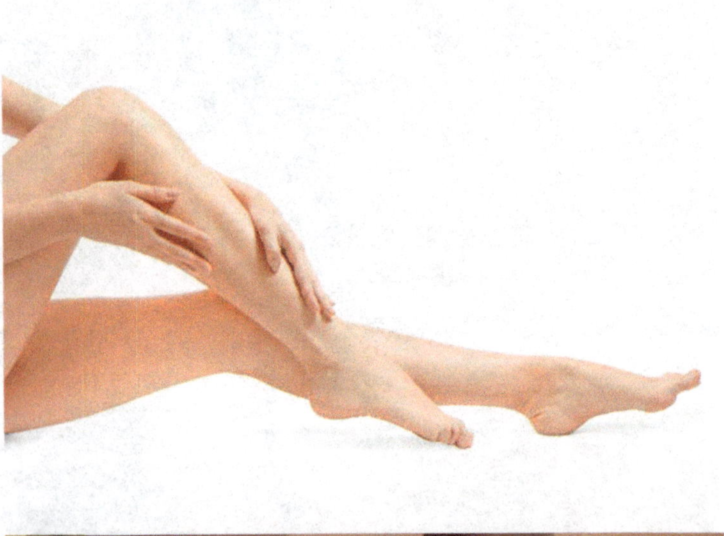

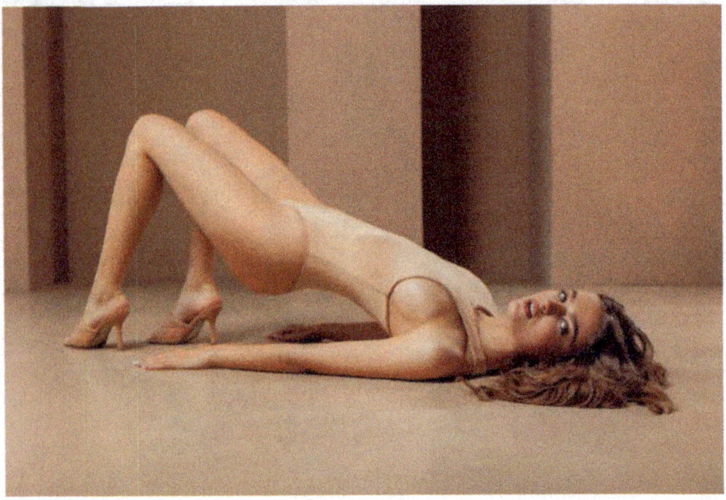

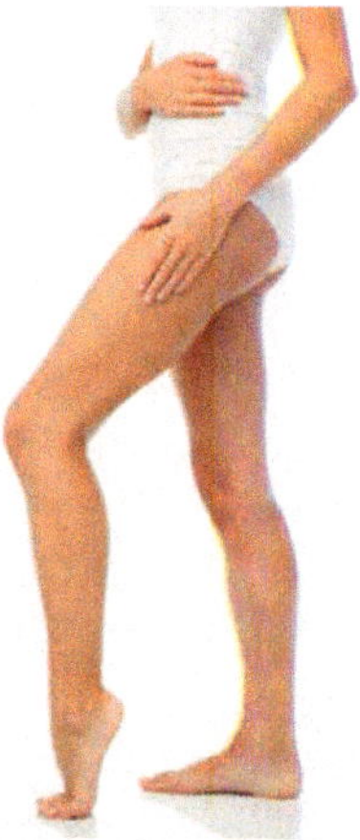

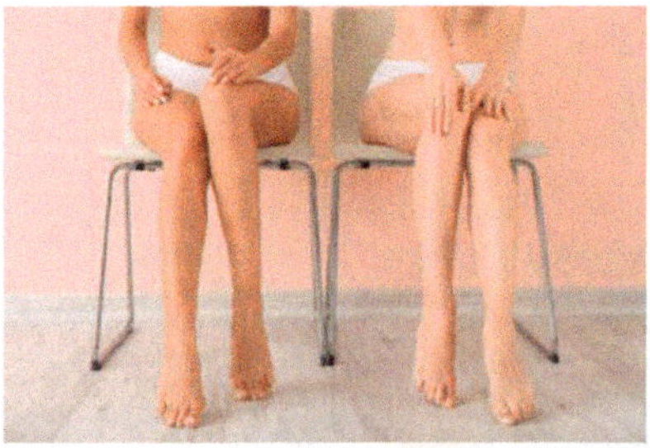

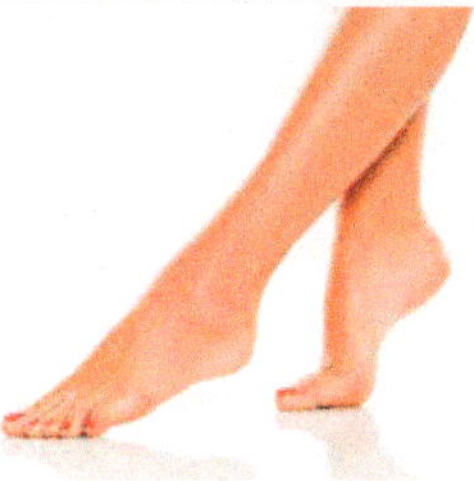

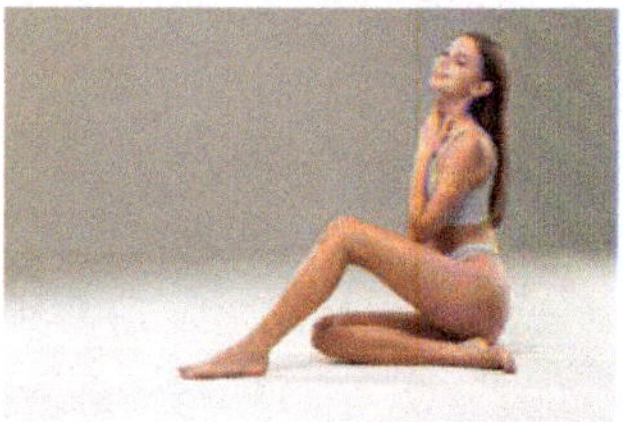

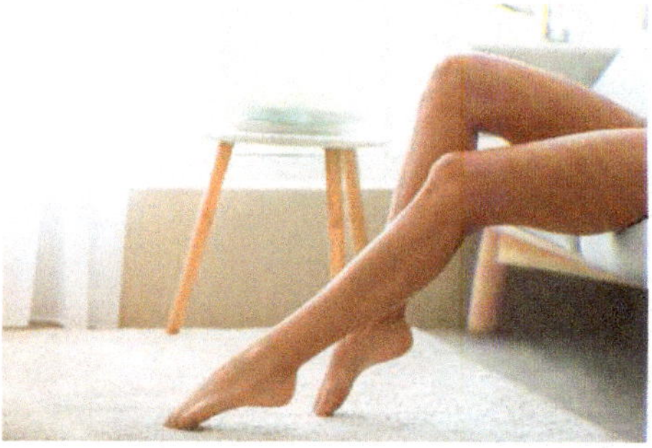

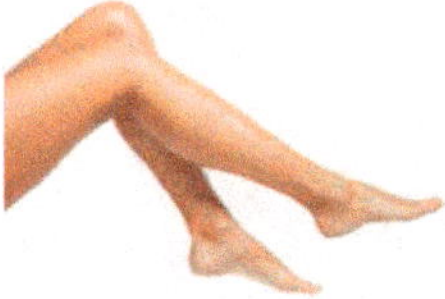

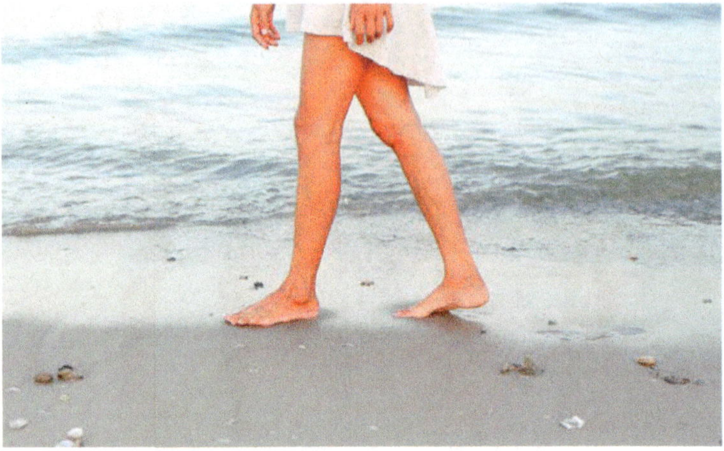

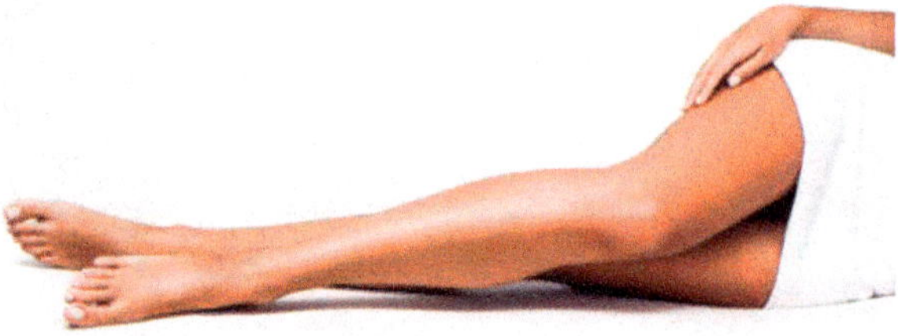

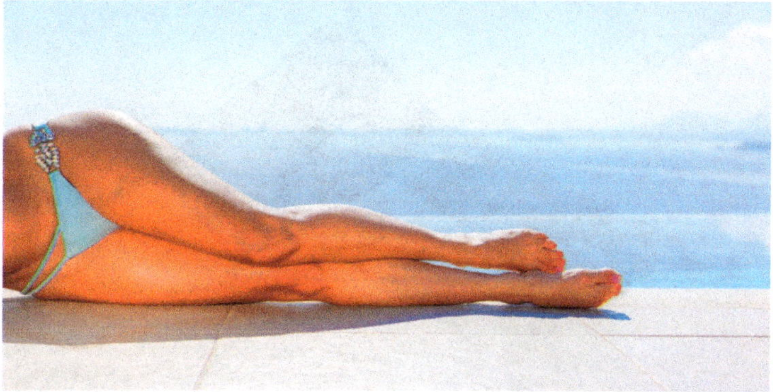

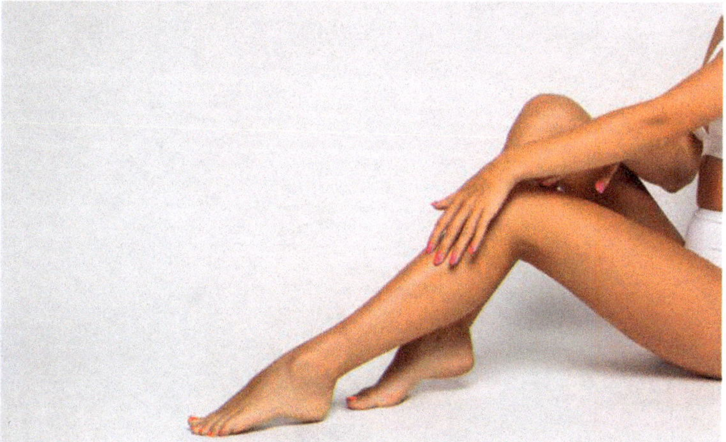

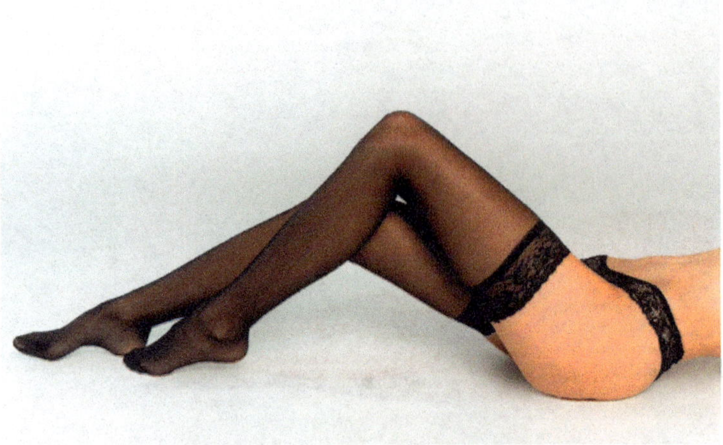

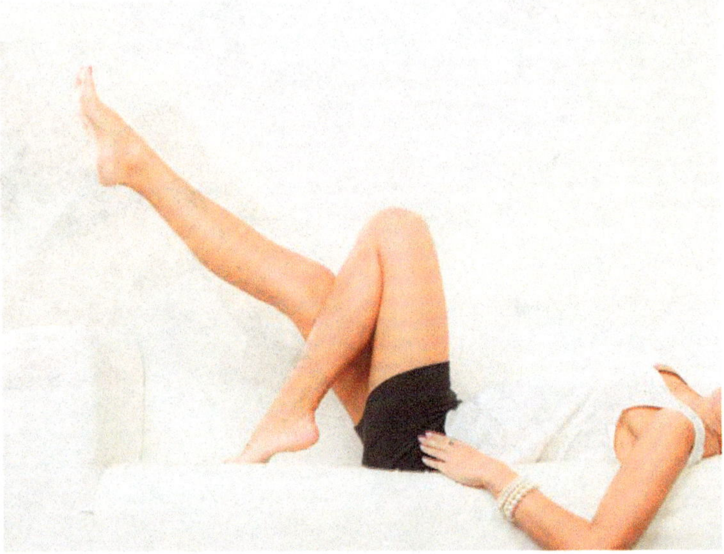

Maybe “Parts” isn’t a perfect subcategory for the analysis of female form respresentations in two dimensions, but, when I was younger, my Aristotian classifying mindset had me collecting notable papers images of ways of respresenting or highlighting particular parts of a woman’s form, so we have the PARTS part of the book The Female Form Frozen, which this series examines in the new light of AI and stock footage.

WAIST

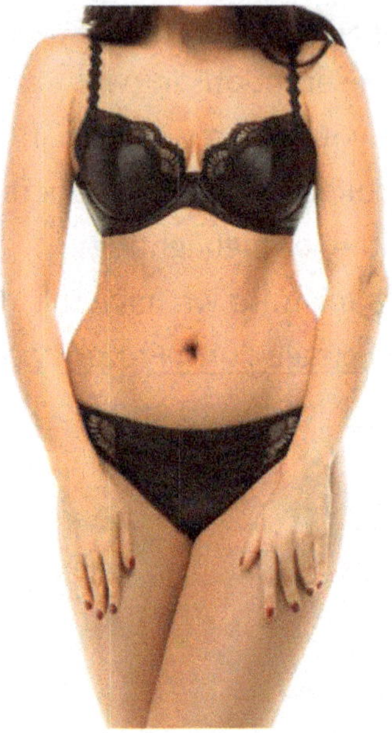

THE FOLLOWING PAGES ARE PAGES FROM THE BOOK "THE FEMALE FORM FROZEN" AND ARE COLLAGES THAT ARE ONE OF A KIND IN ORIGINAL PAPER AND COLOR FOR SALE FOR 1500 DOLLARS A PIECE

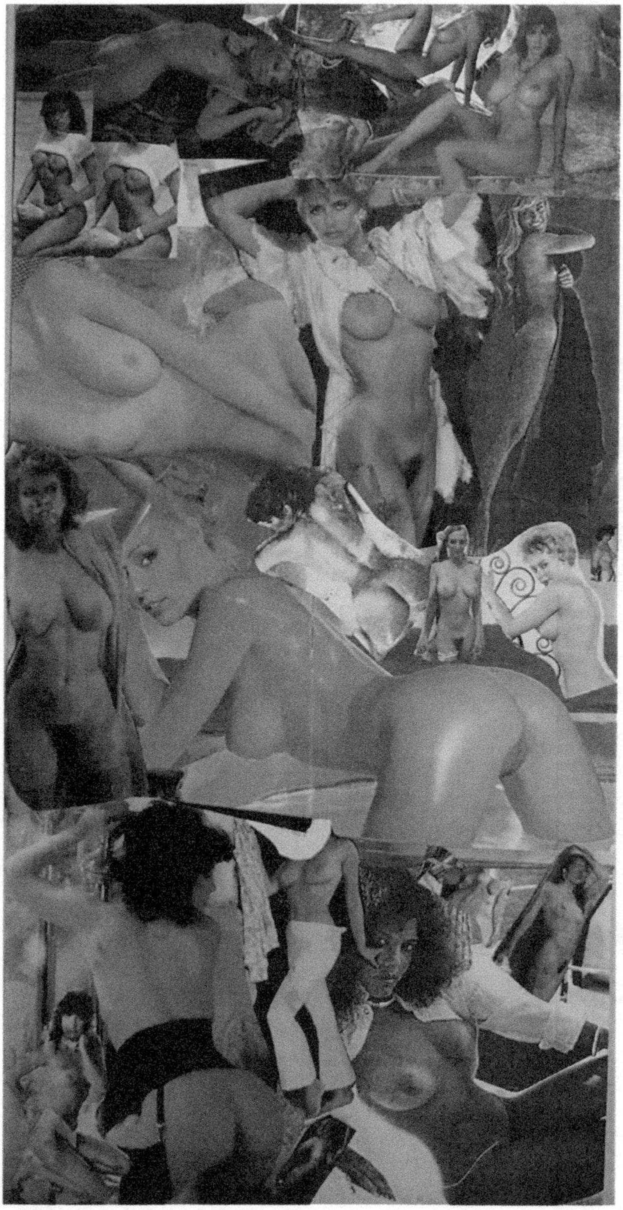

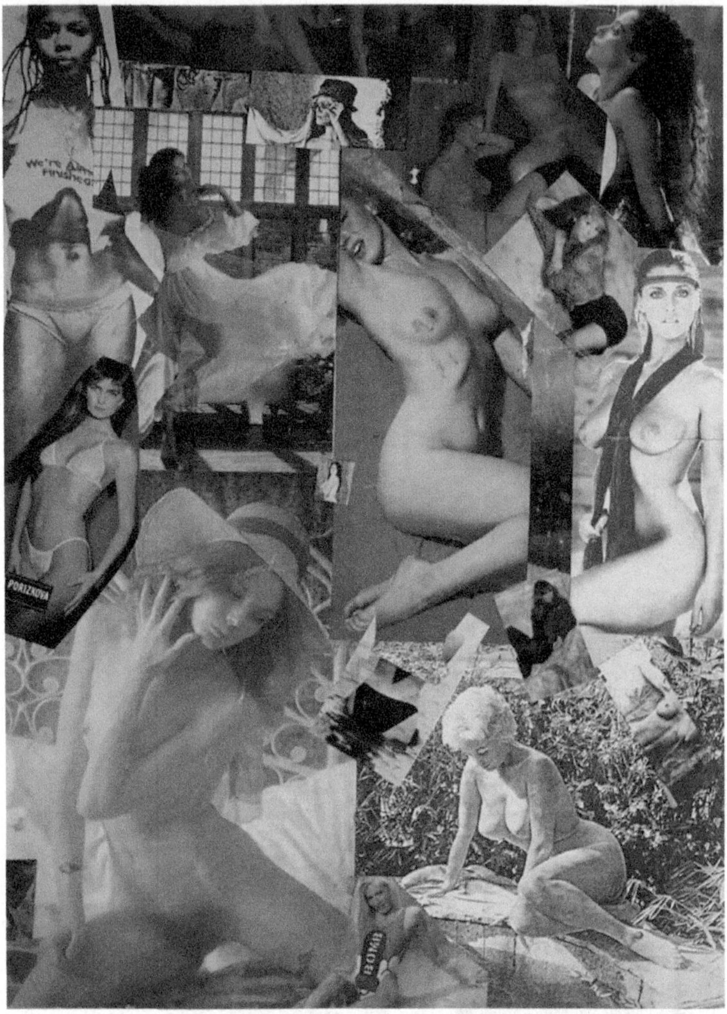
PORIZKOVA

BELLIES -- (CAN YOU TELL WHICH ONES ARE STOCK AND WHICH ONES ARE AI?)

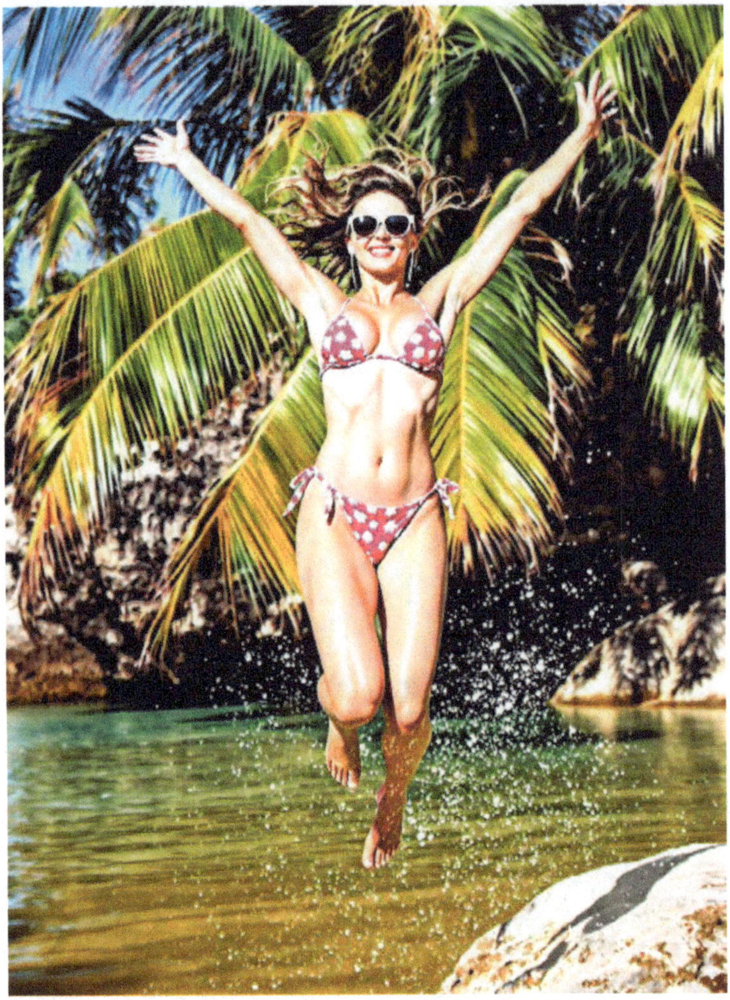

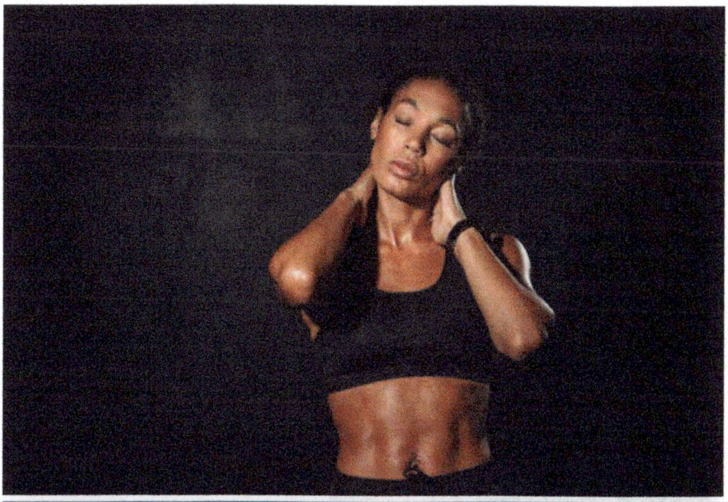

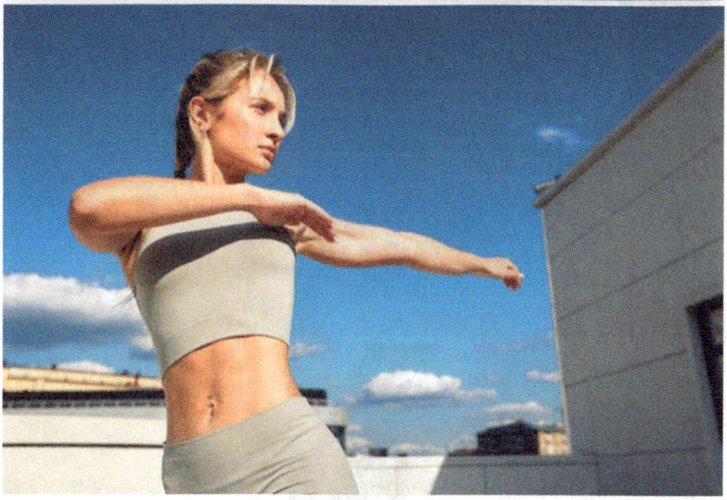

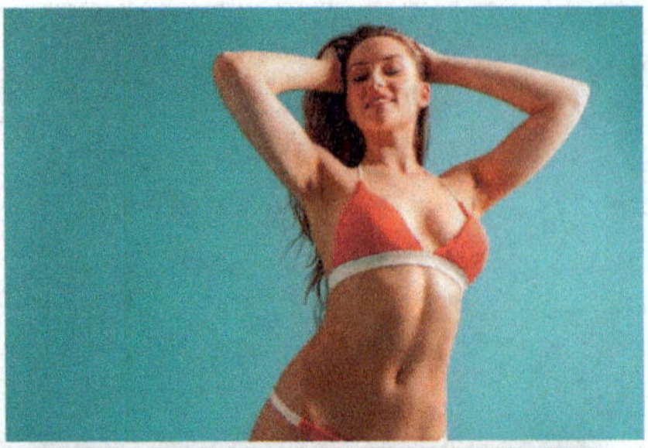

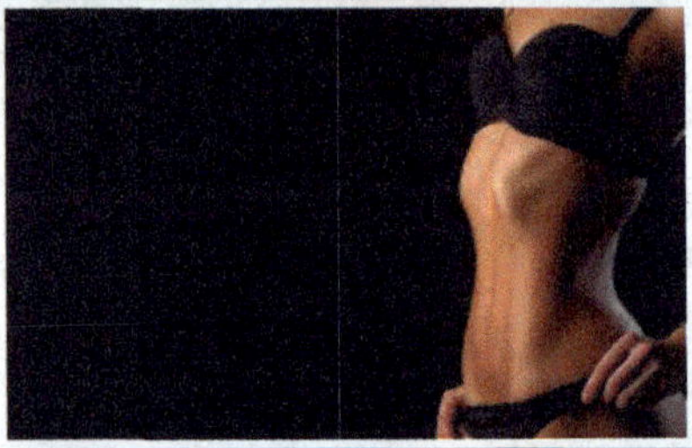

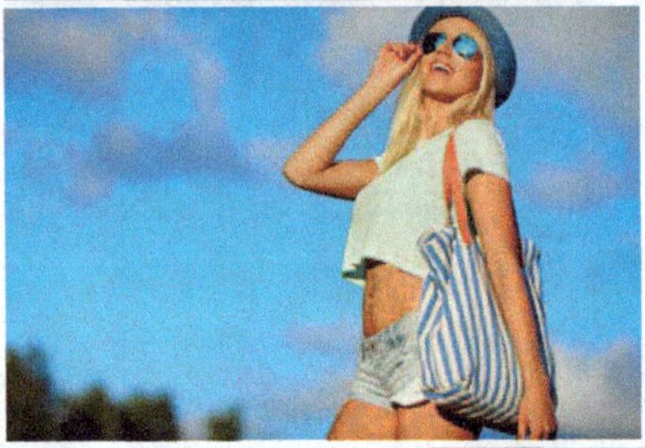

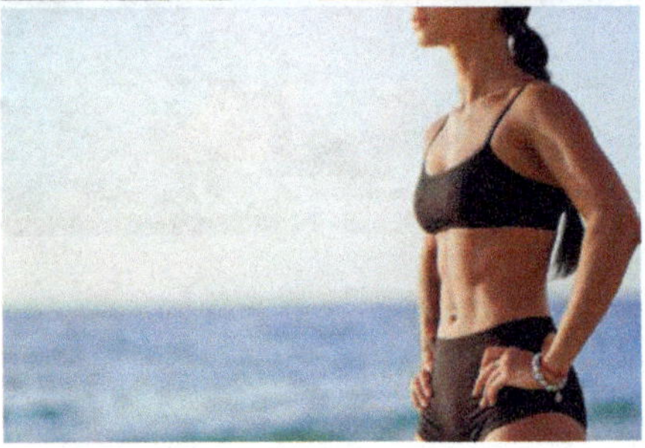

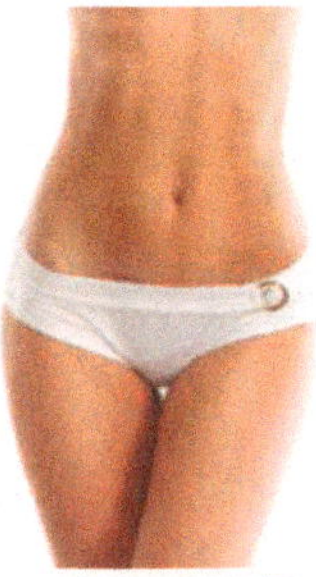
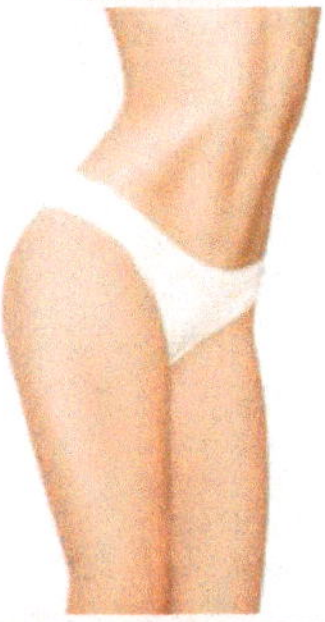
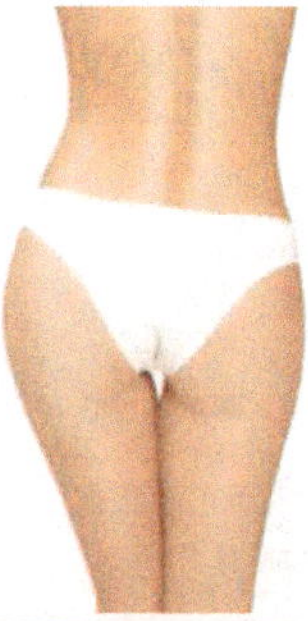
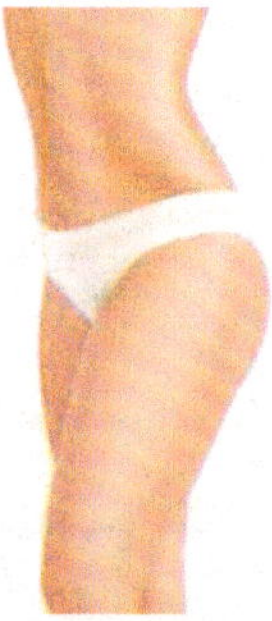
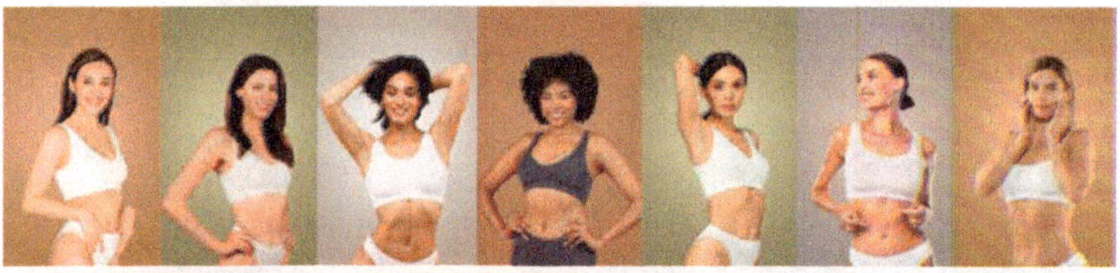
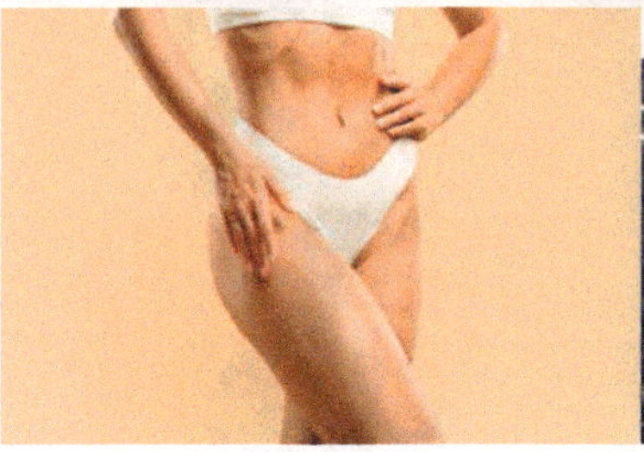
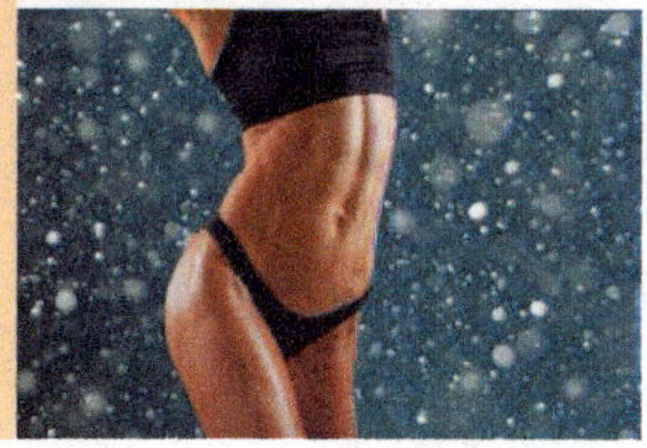

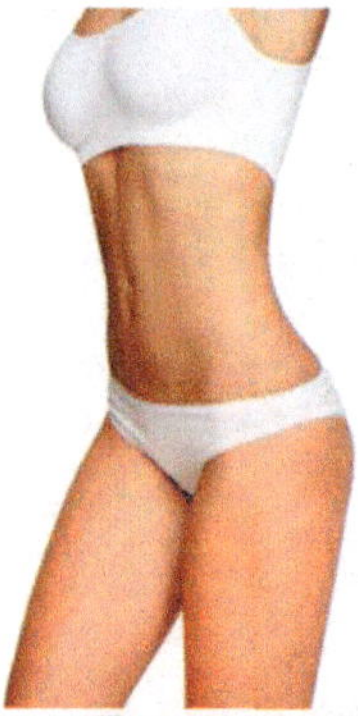
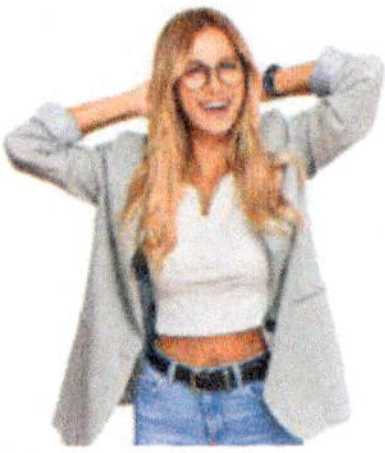
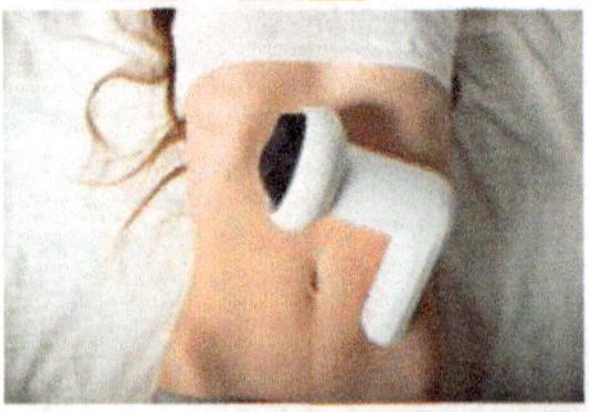
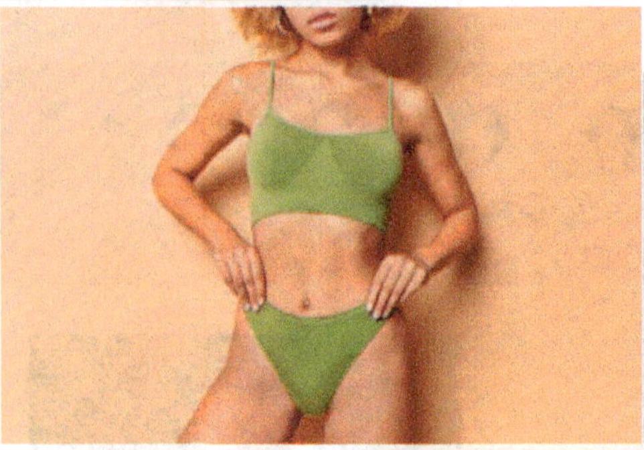
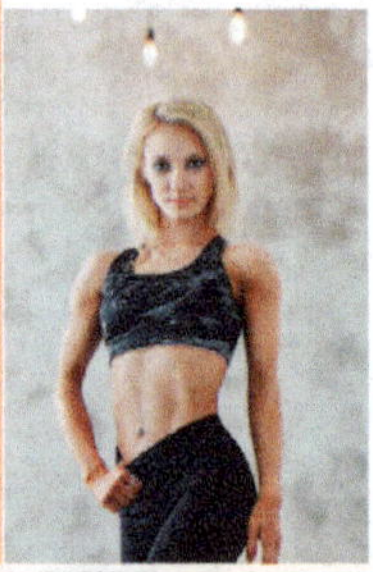
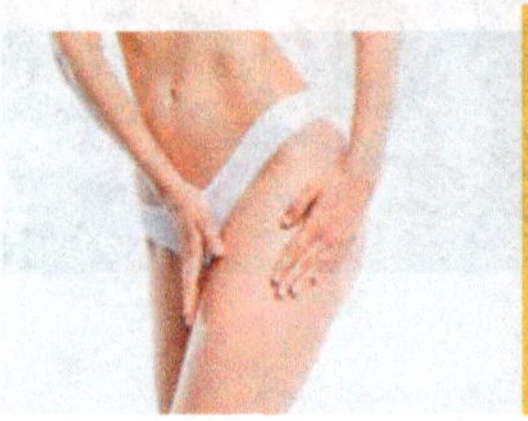
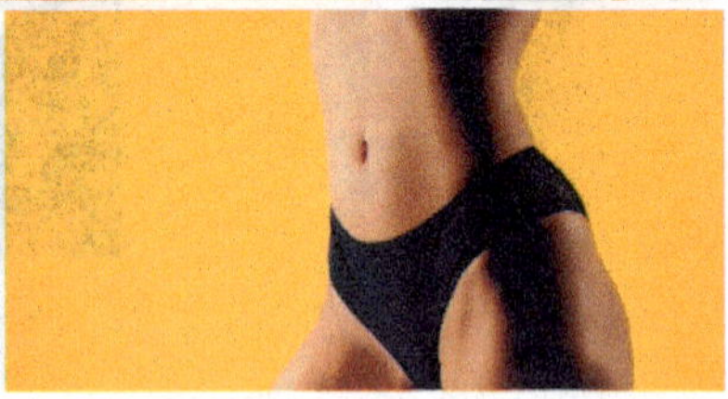

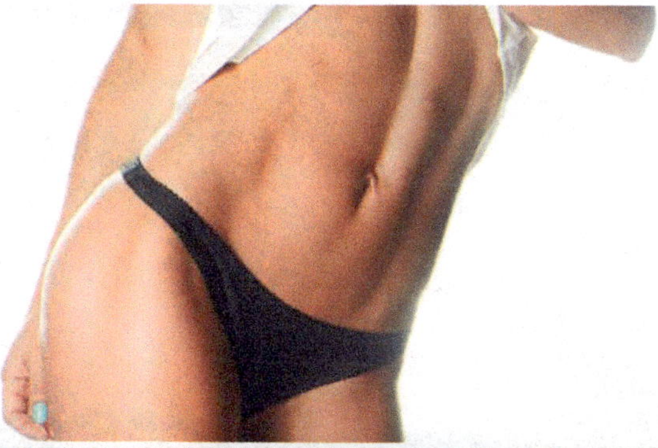

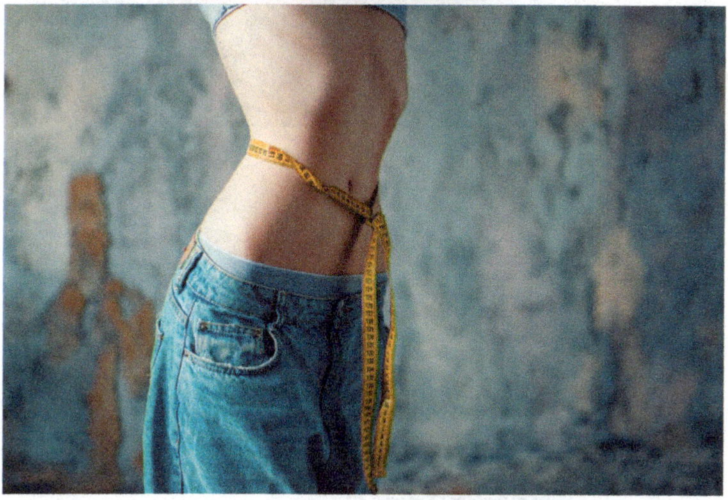

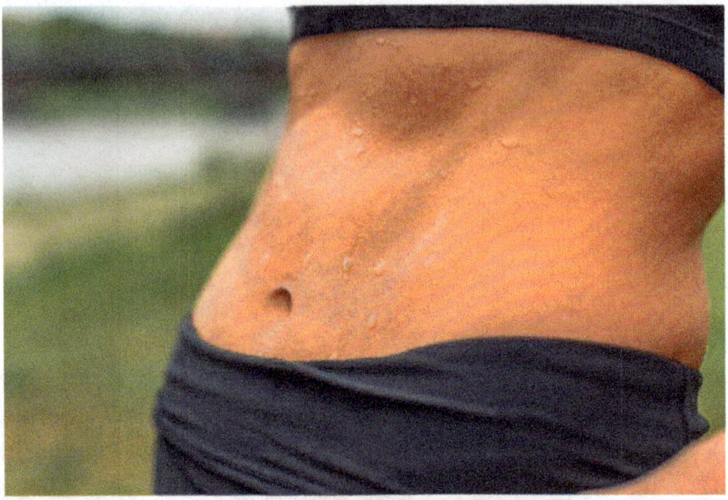

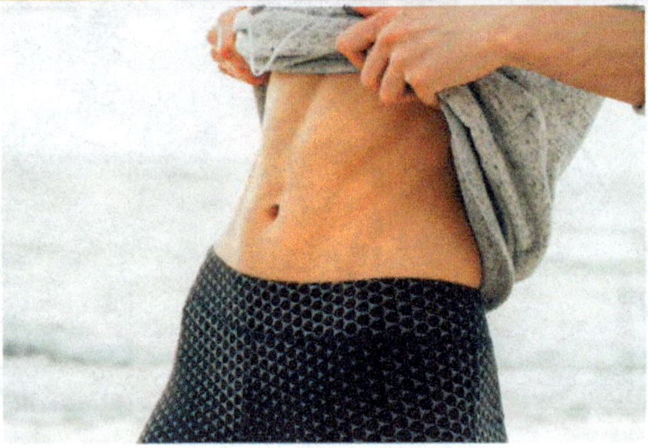

THE FOLLOWING PAGES ARE PAGES FROM THE BOOK "THE FEMALE FORM FROZEN" AND ARE COLLAGES THAT ARE ONE OF A KIND IN ORIGINAL PAPER AND COLOR FOR SALE FOR 1500 DOLLARS A PIECE

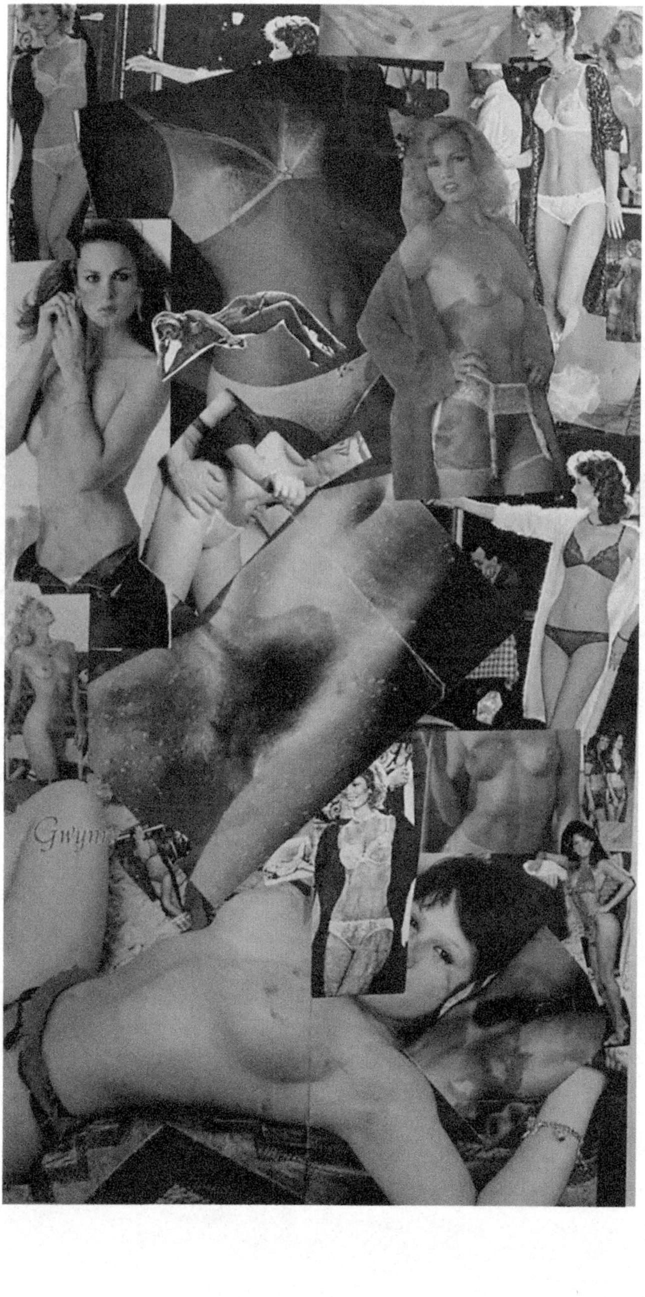

BACK ARCHED ETC

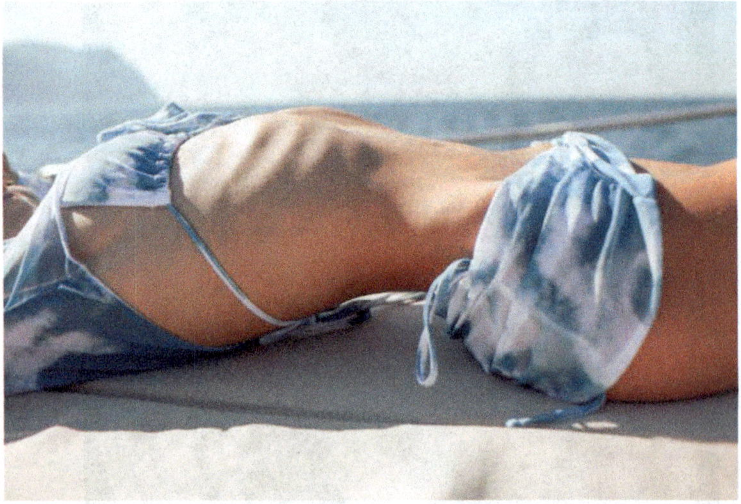

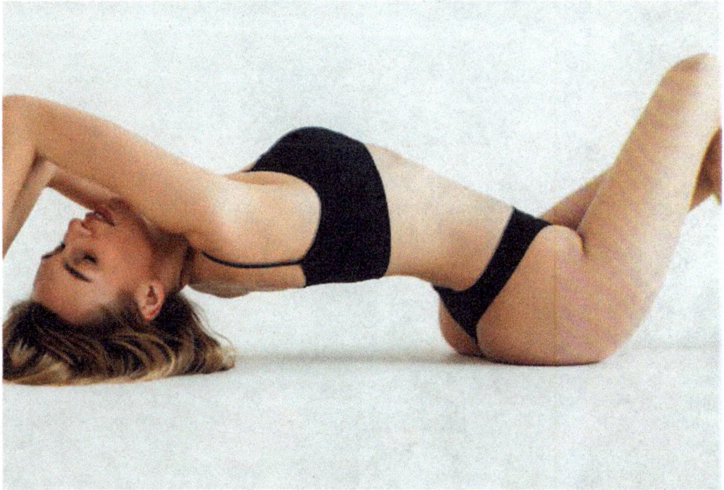

THE FOLLOWING PAGES ARE PAGES FROM THE BOOK "THE FEMALE FORM FROZEN" AND ARE COLLAGES THAT ARE ONE OF A KIND IN ORIGINAL PAPER AND COLOR FOR SALE FOR 1500 DOLLARS A PIECE

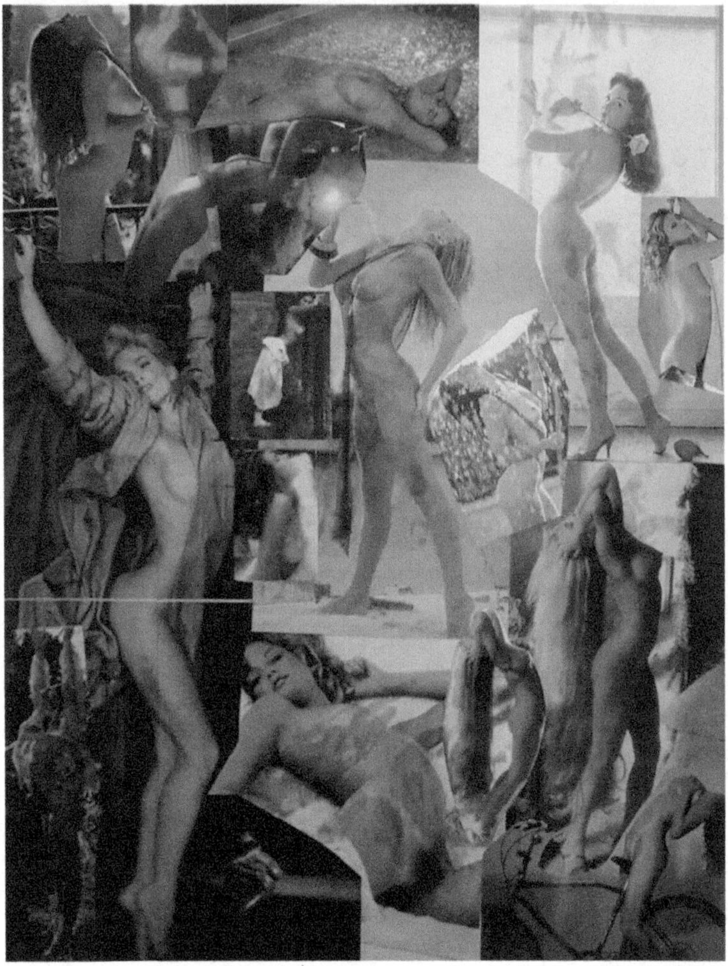

BACKS

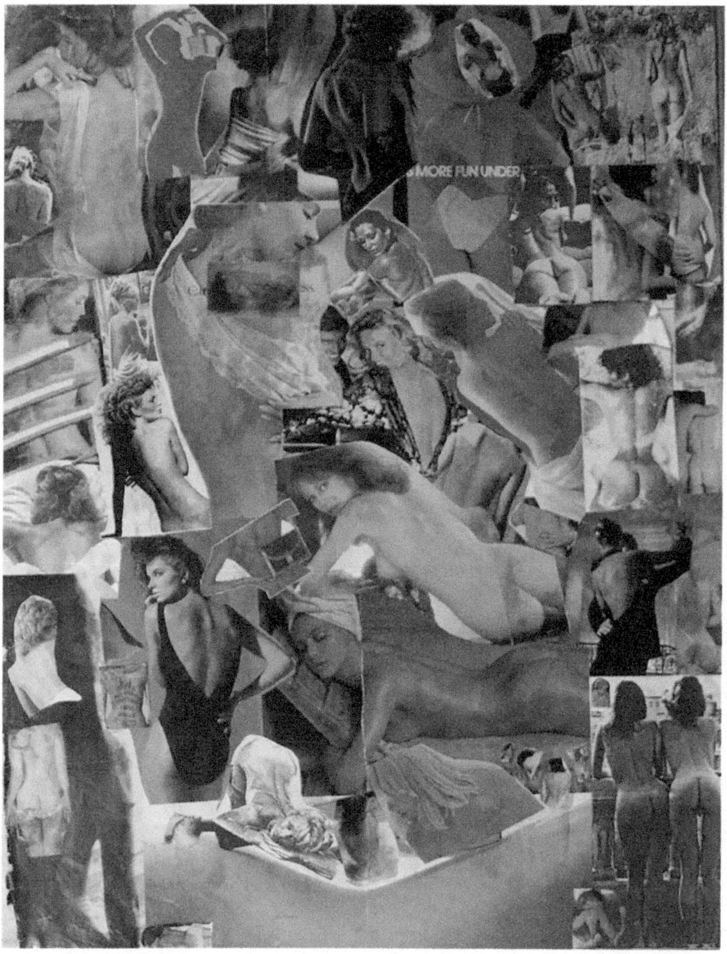

HANDS AND FEET STOCK photos

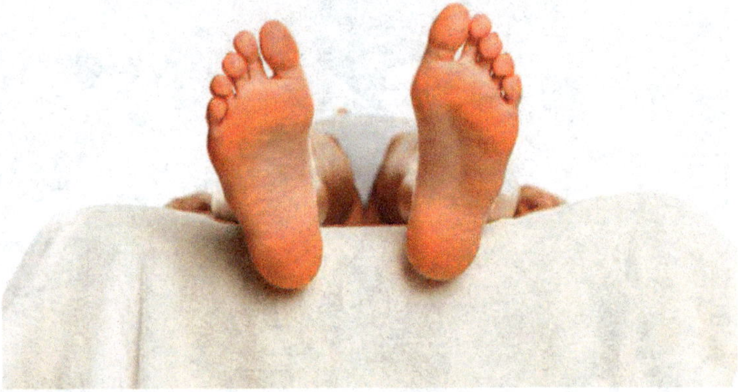

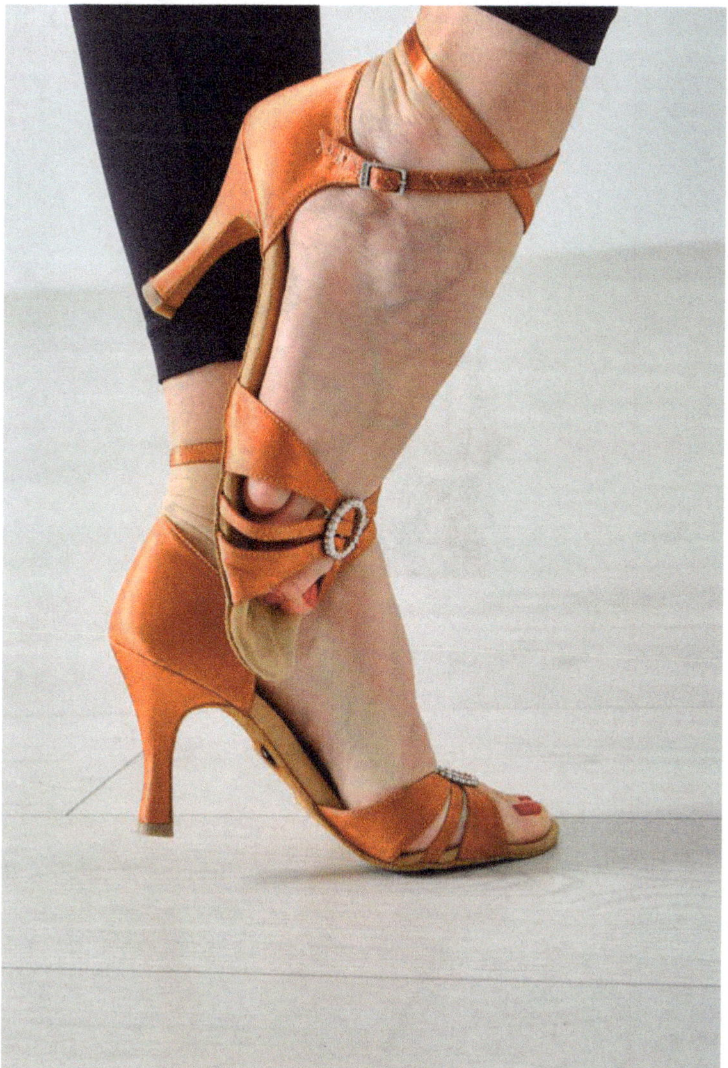

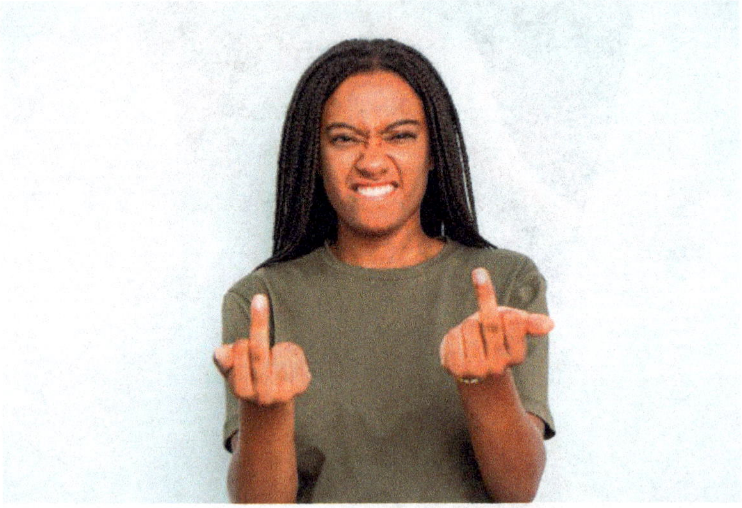

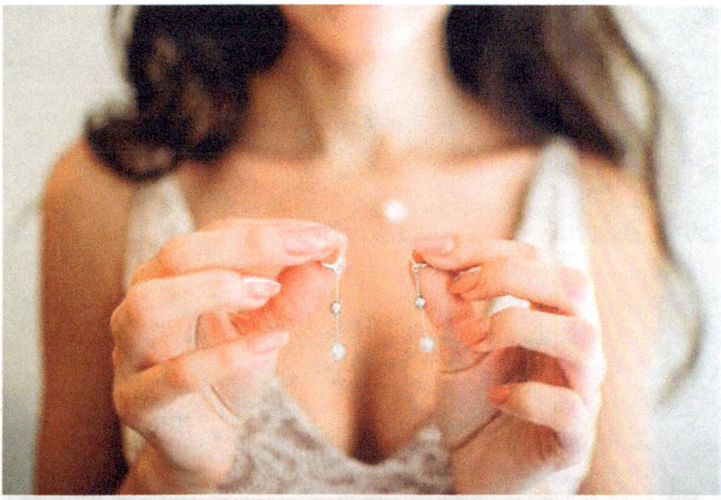

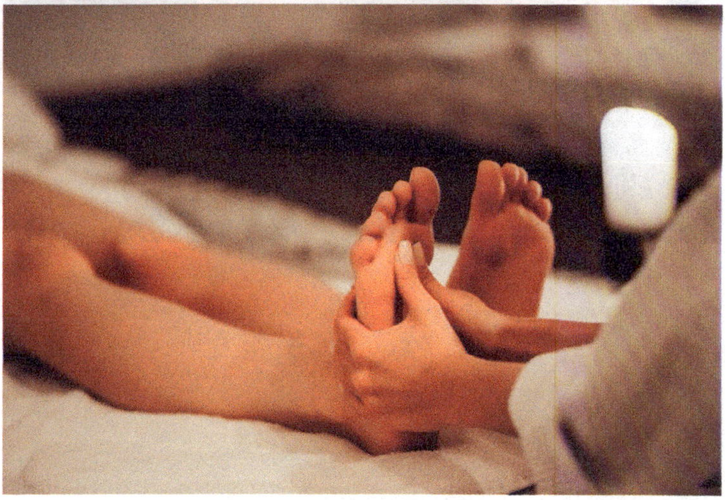

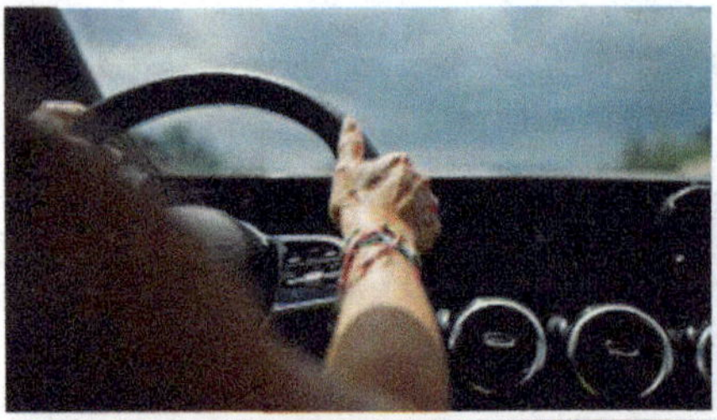
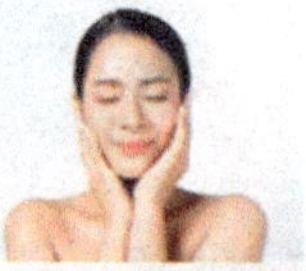
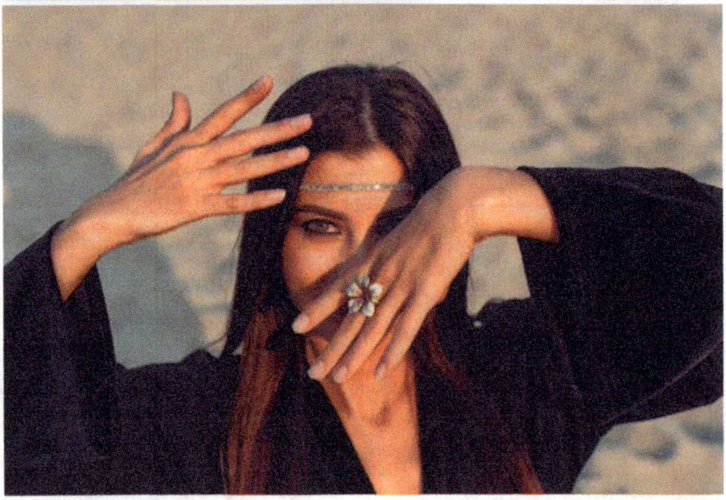
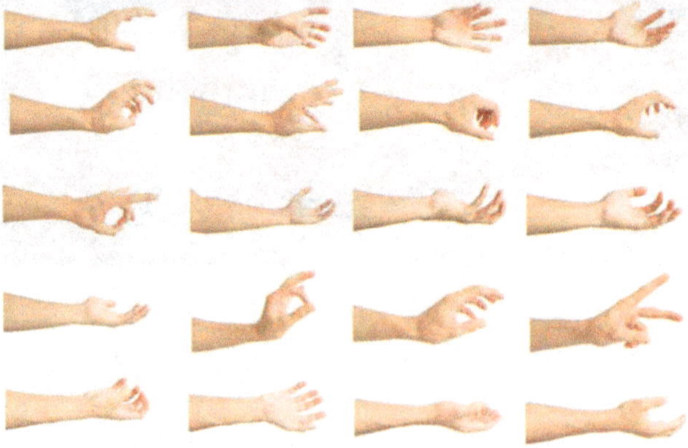

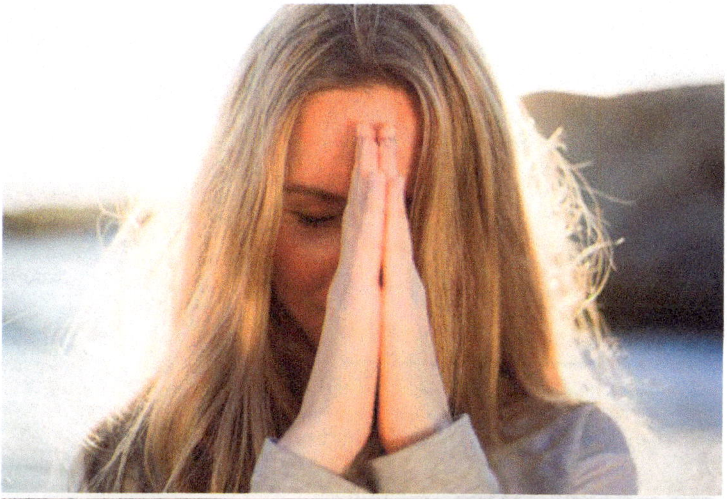

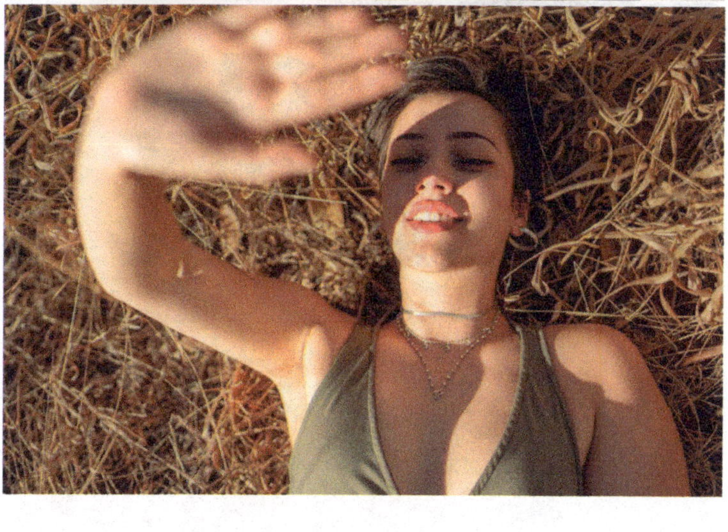

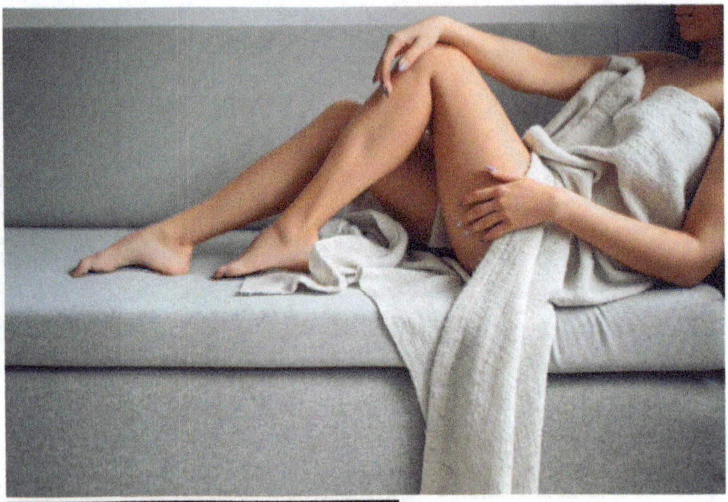

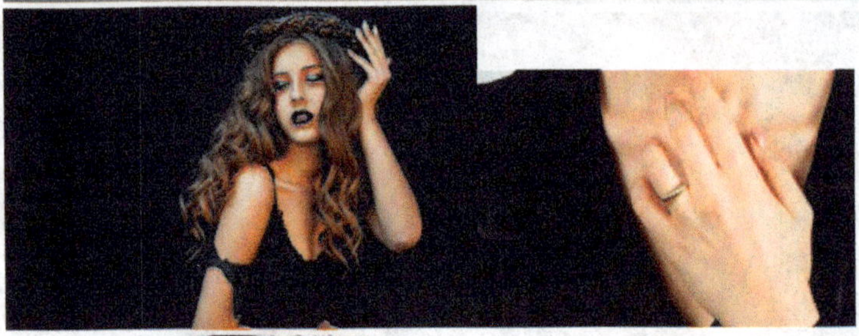

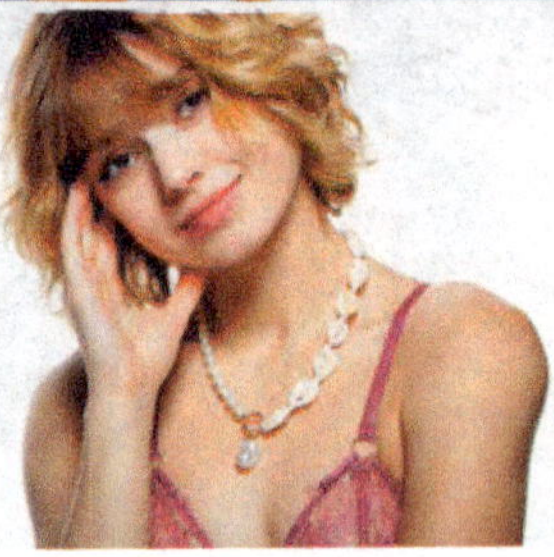

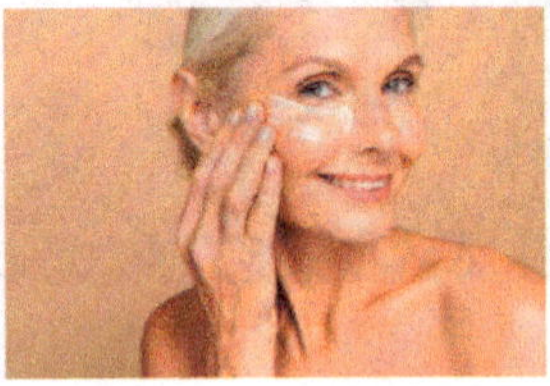

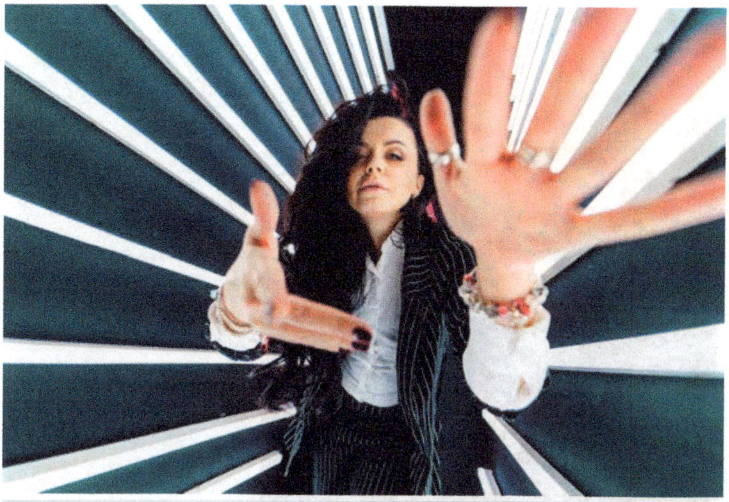

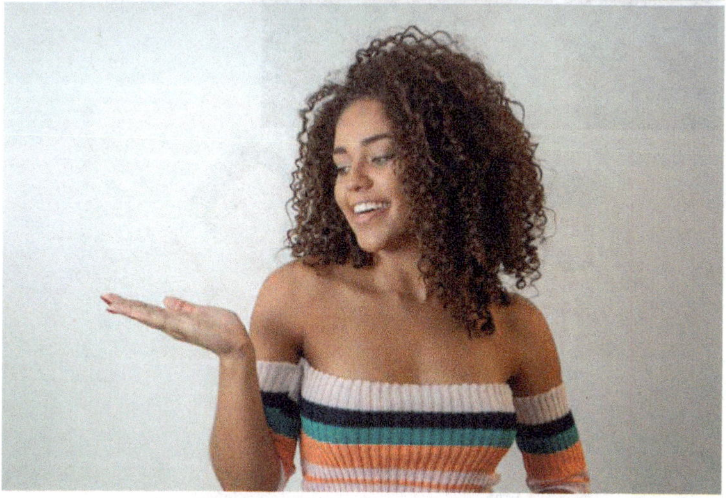

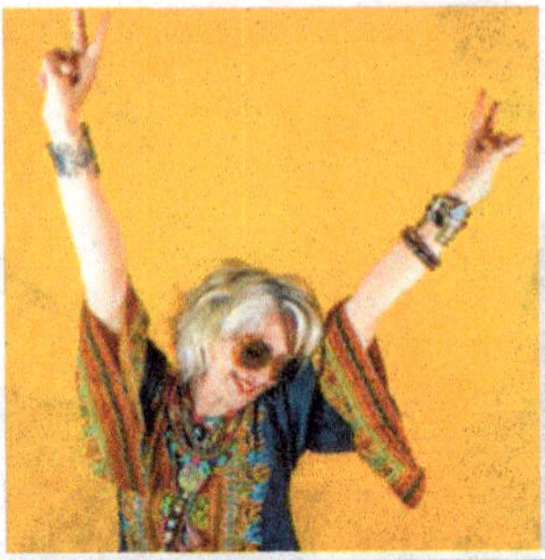
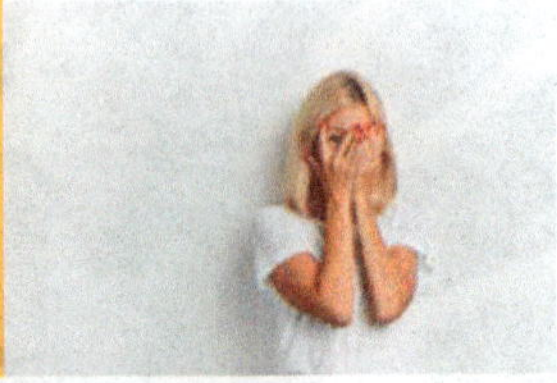
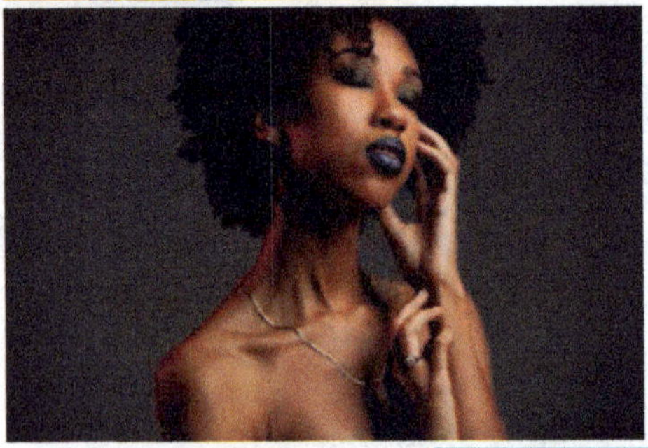
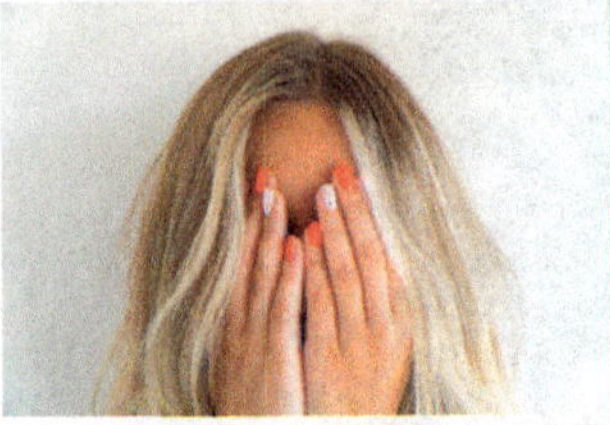
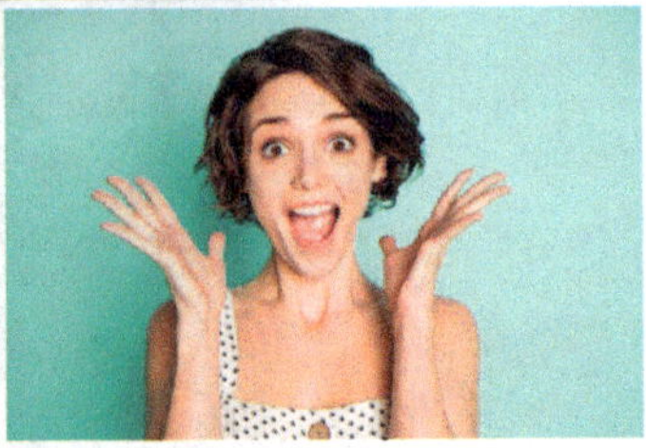
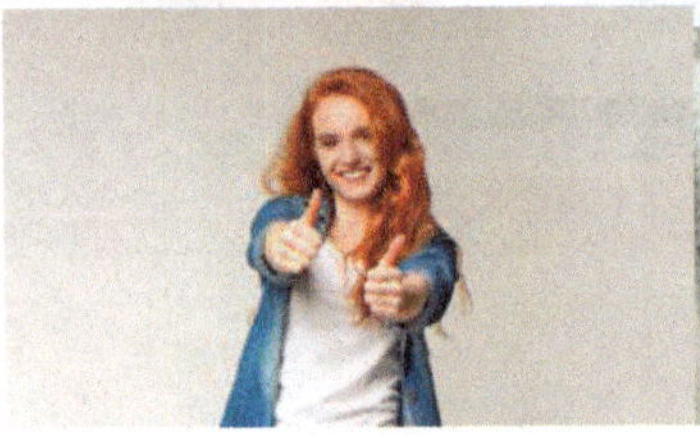
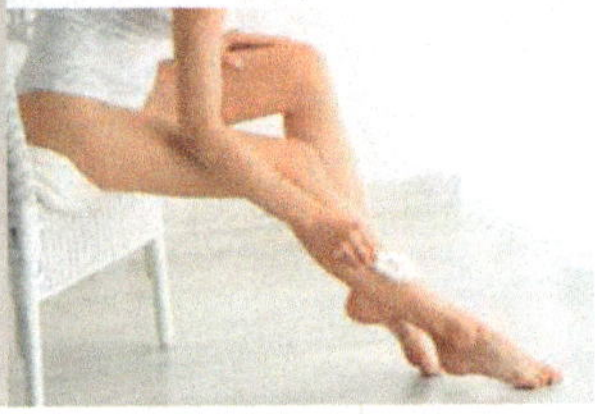

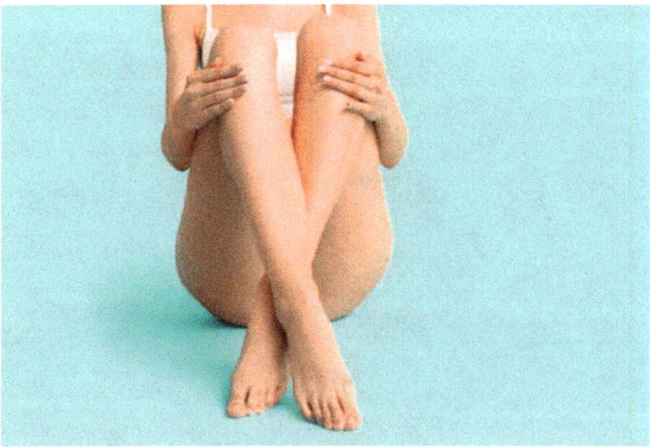

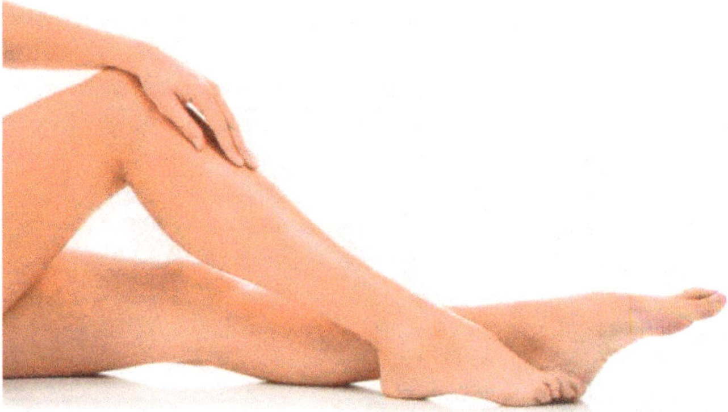

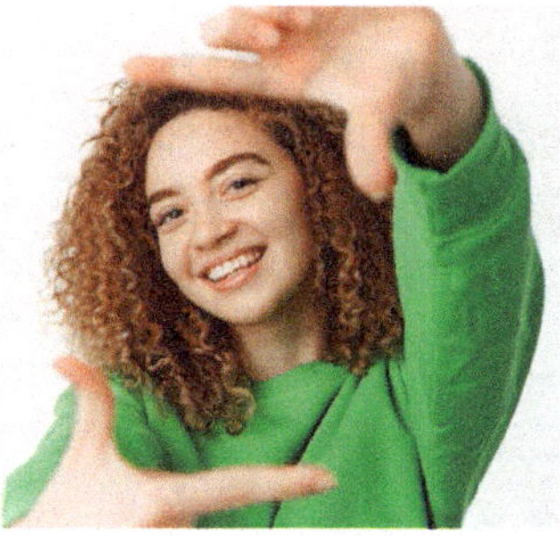

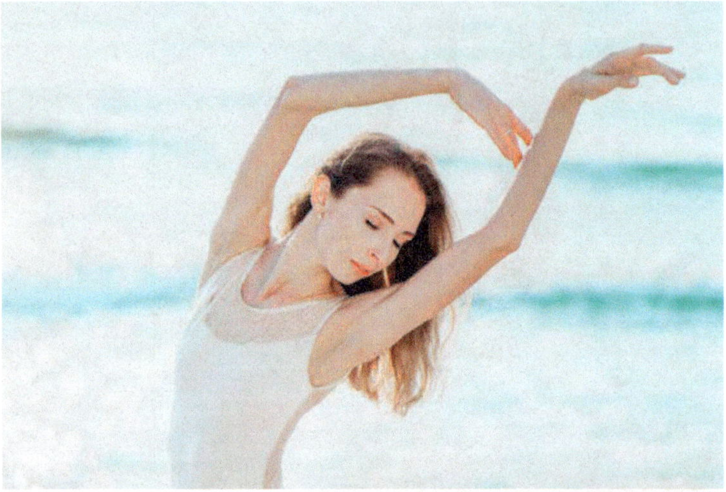

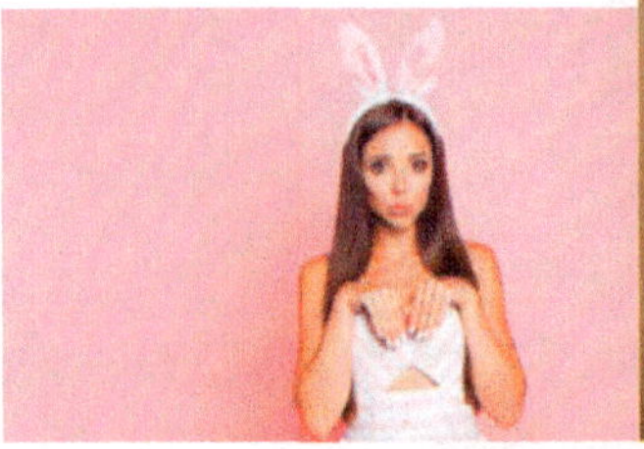

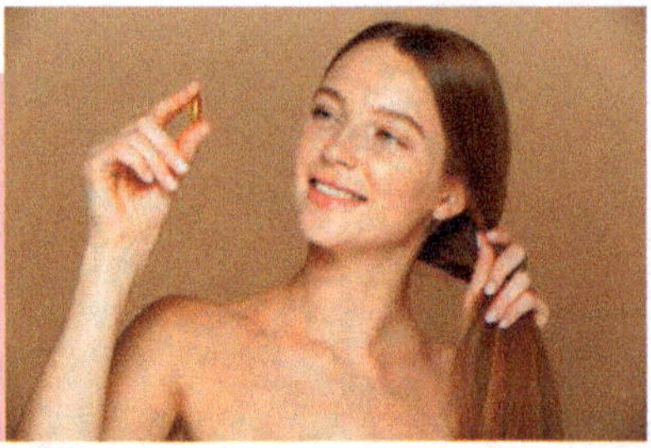

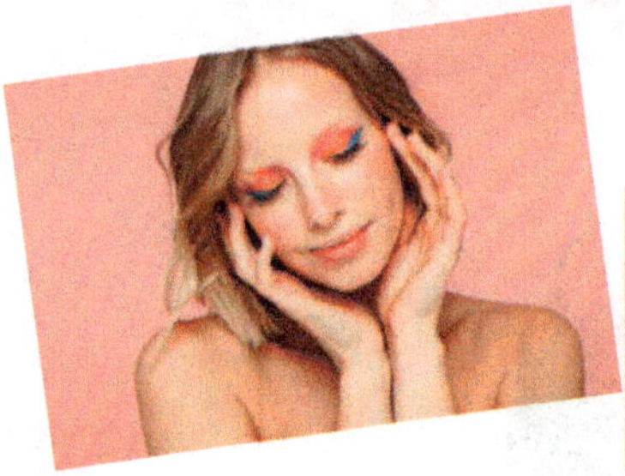

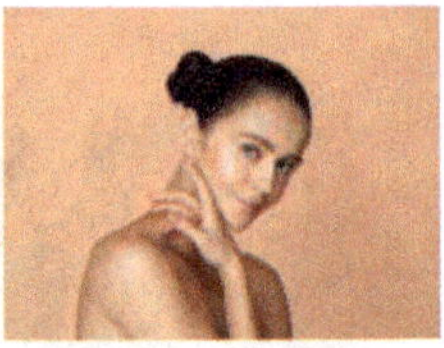

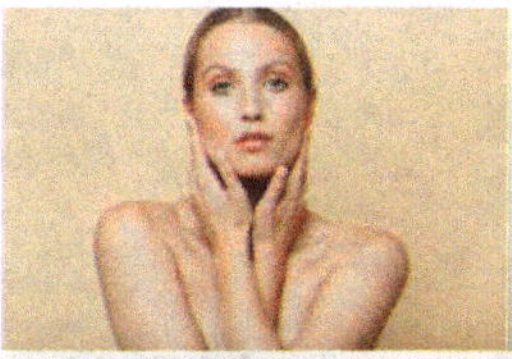
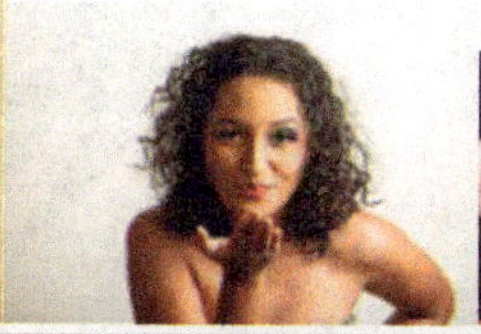
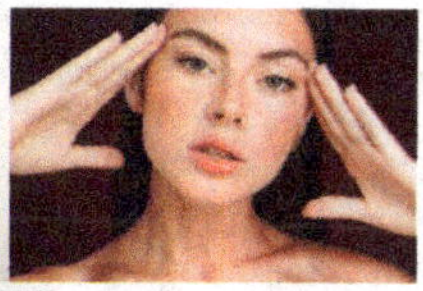
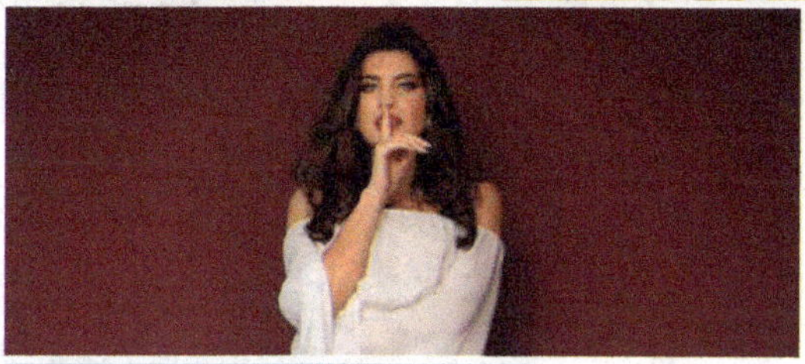
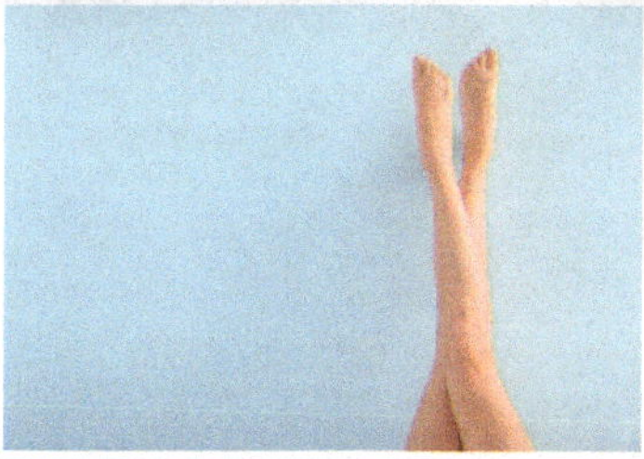
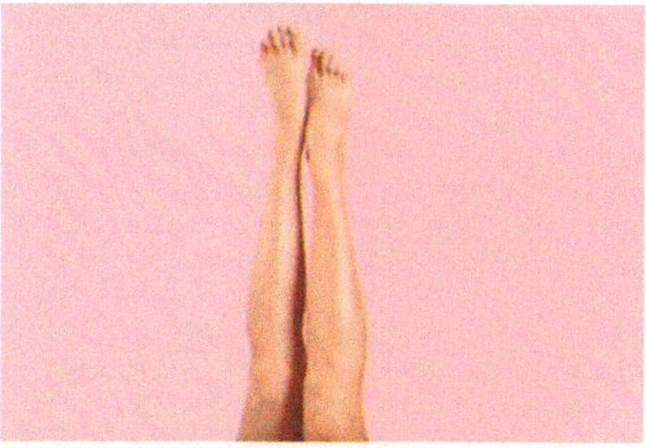
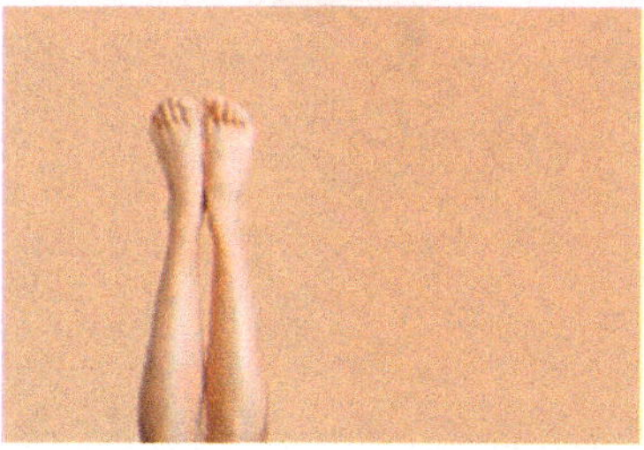

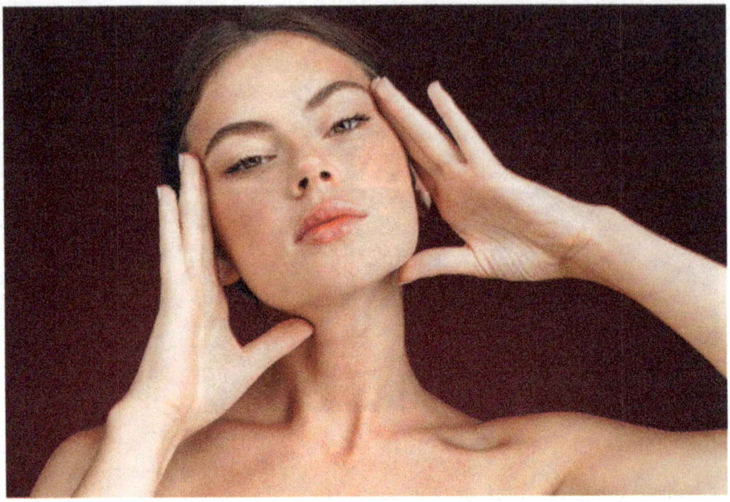

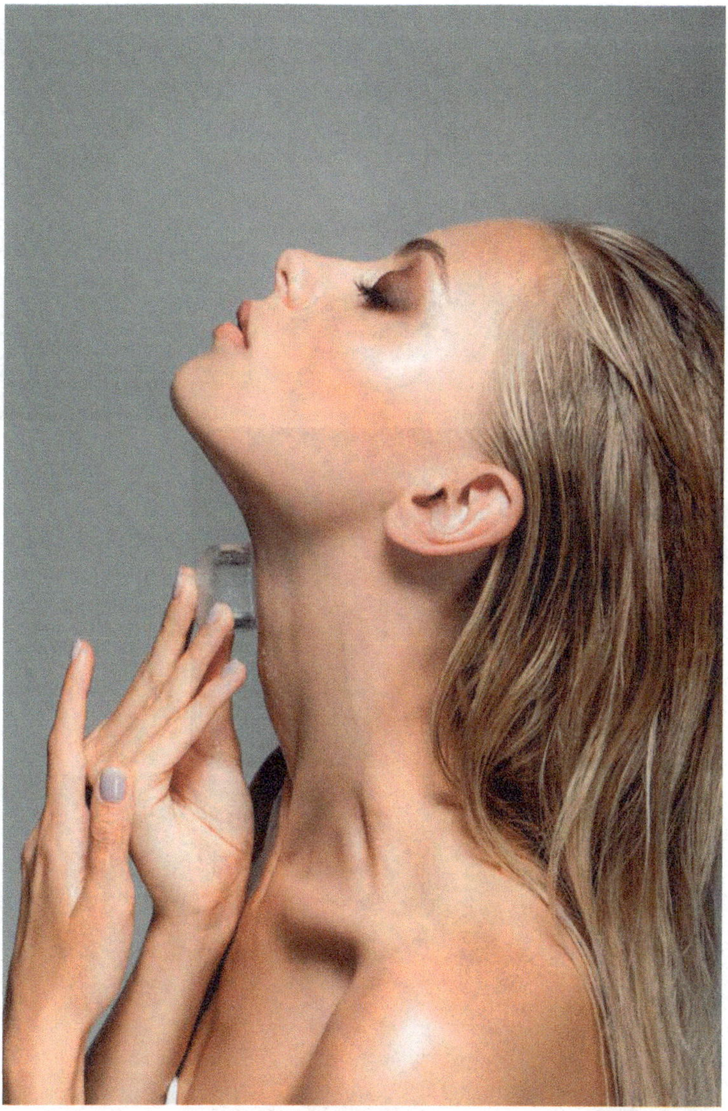

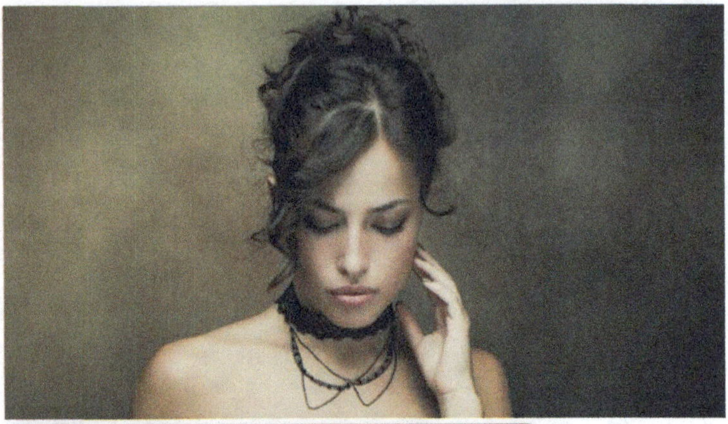

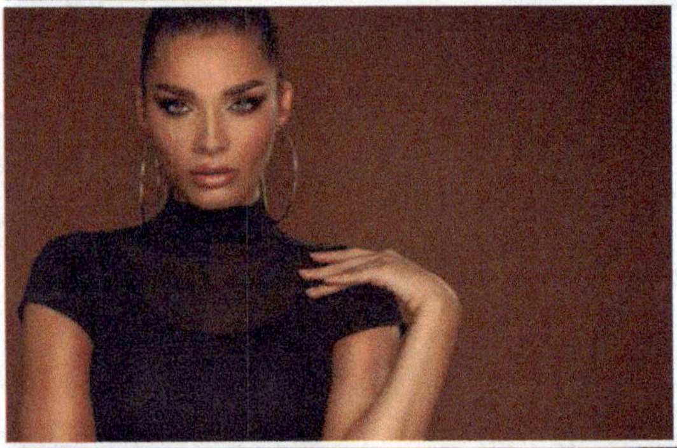

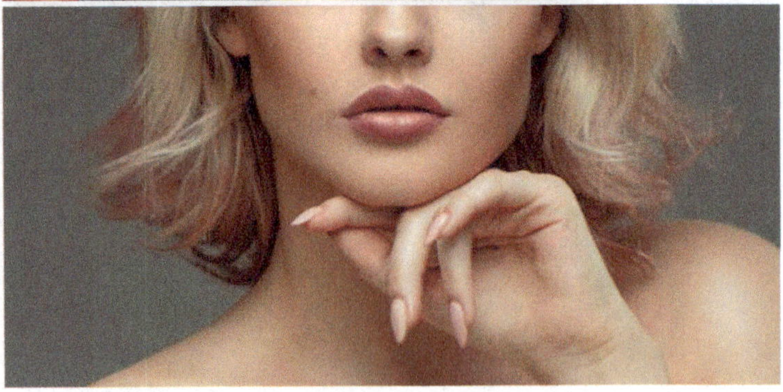

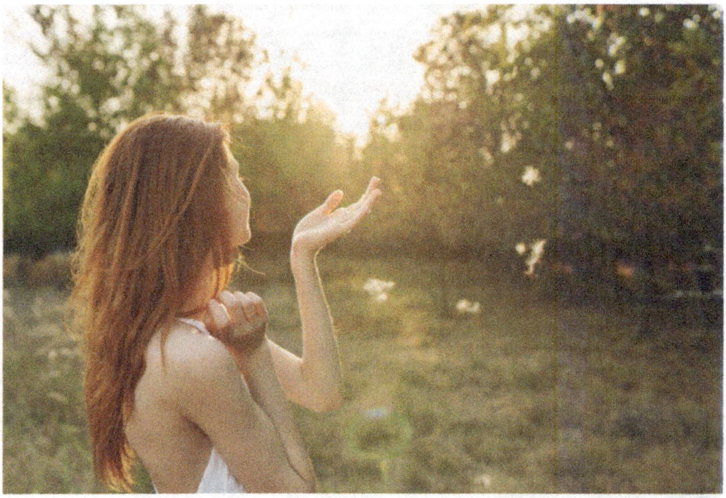

THE FOLLOWING PAGES ARE PAGES FROM THE BOOK "THE FEMALE FORM FROZEN" AND ARE COLLAGES THAT ARE ONE OF A KIND IN ORIGINAL PAPER AND COLOR FOR SALE FOR 1500 DOLLARS A PIECE

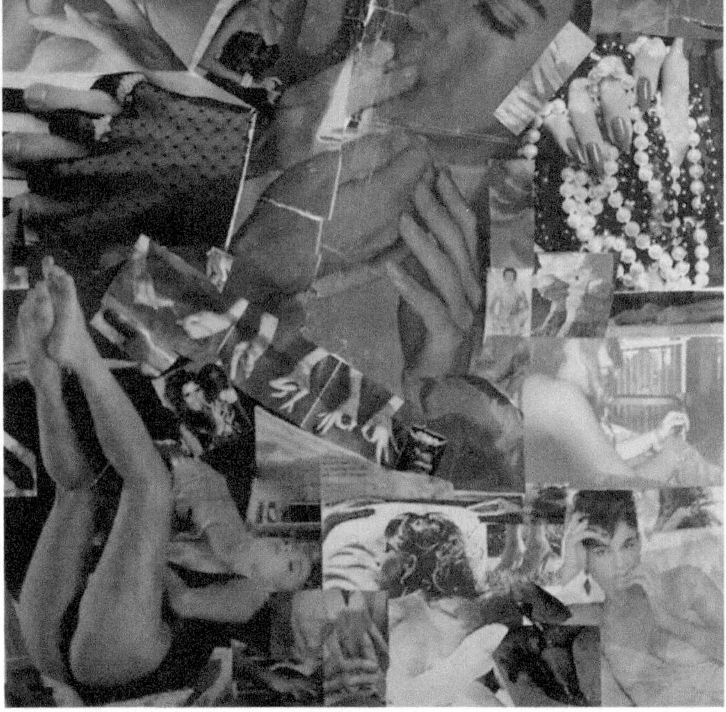

TONGUES

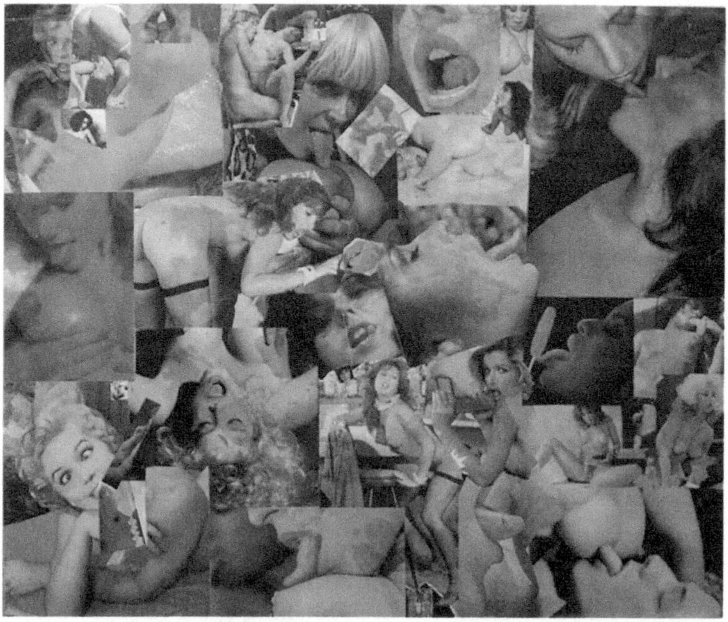

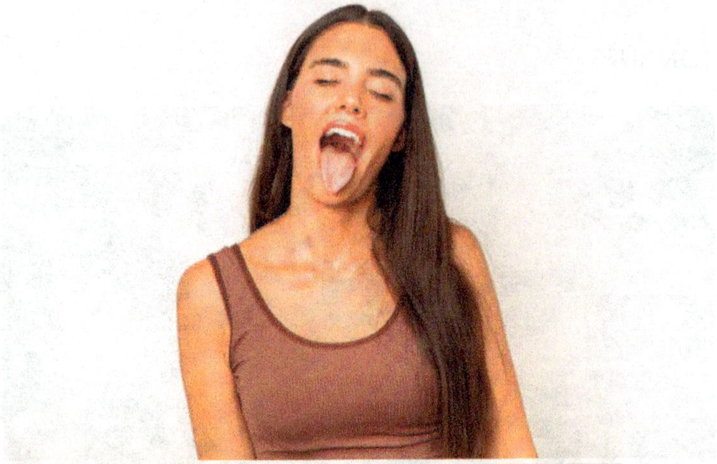

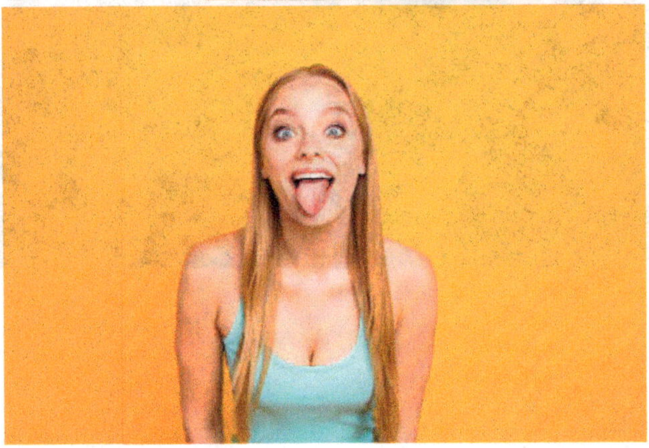

THE FOLLOWING PAGES ARE PAGES FROM THE BOOK "THE FEMALE FORM FROZEN" AND ARE COLLAGES THAT ARE ONE OF A KIND IN ORIGINAL PAPER AND COLOR FOR SALE FOR 1500 DOLLARS A PIECE

SHOULDERS AND COLLAR BONES

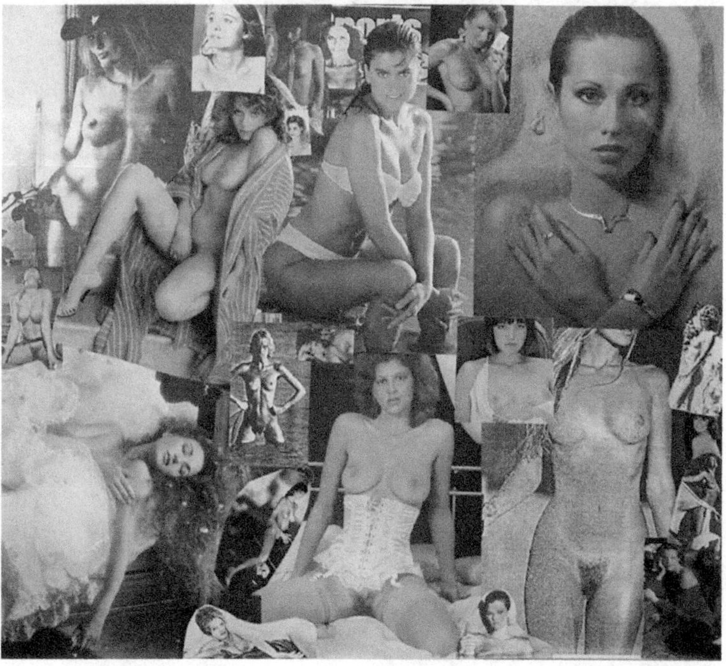

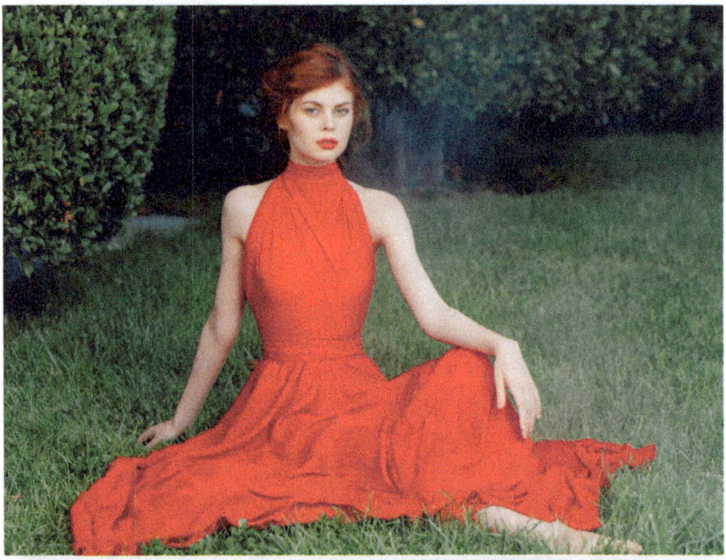

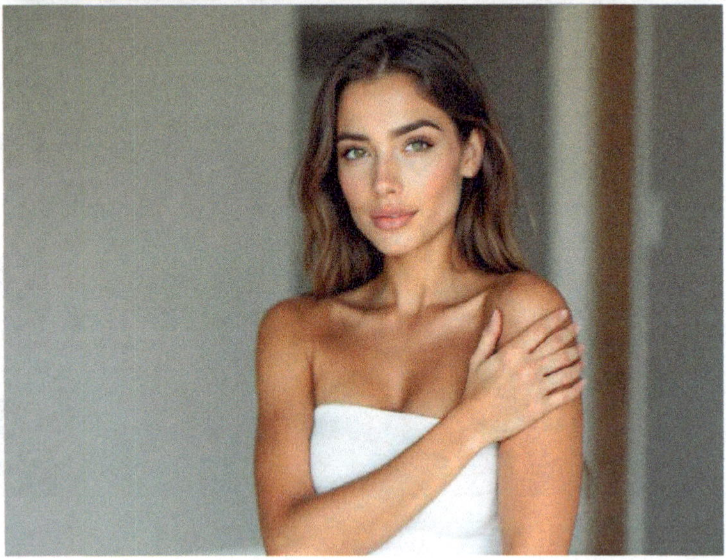

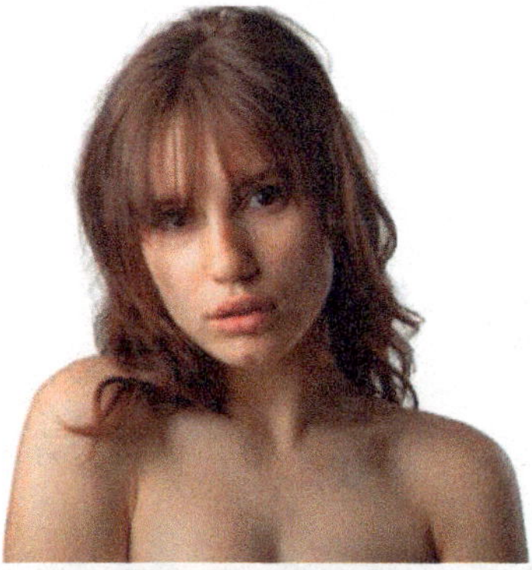

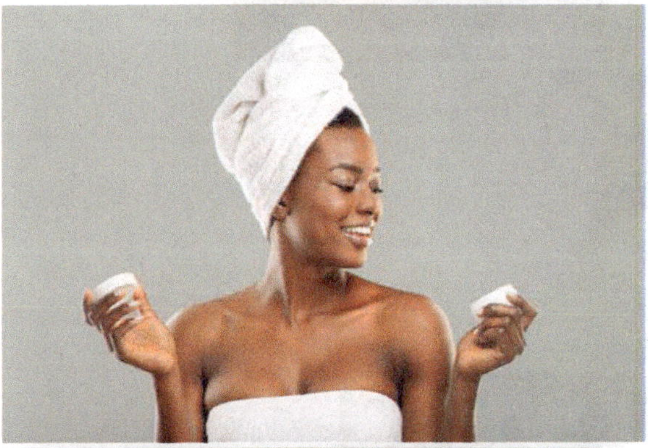

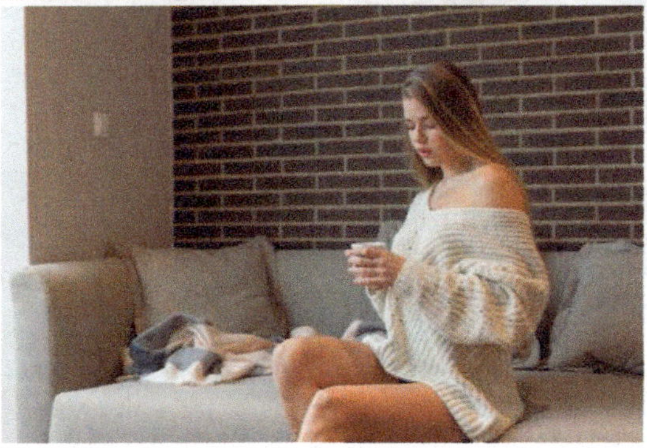

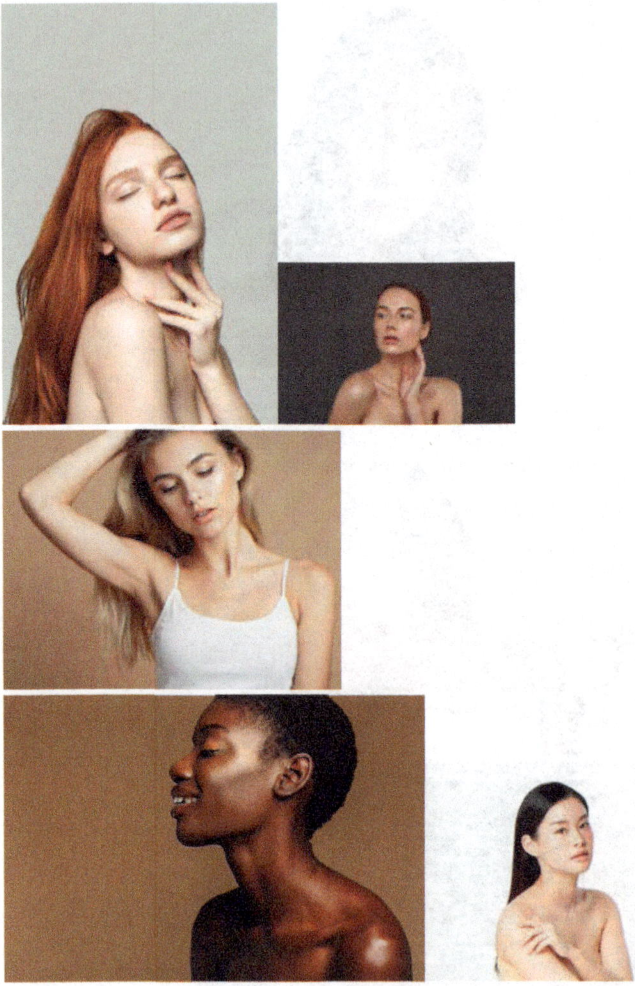

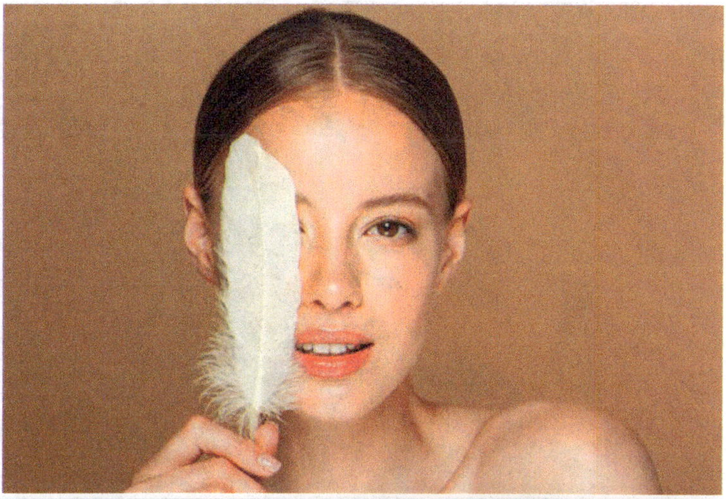

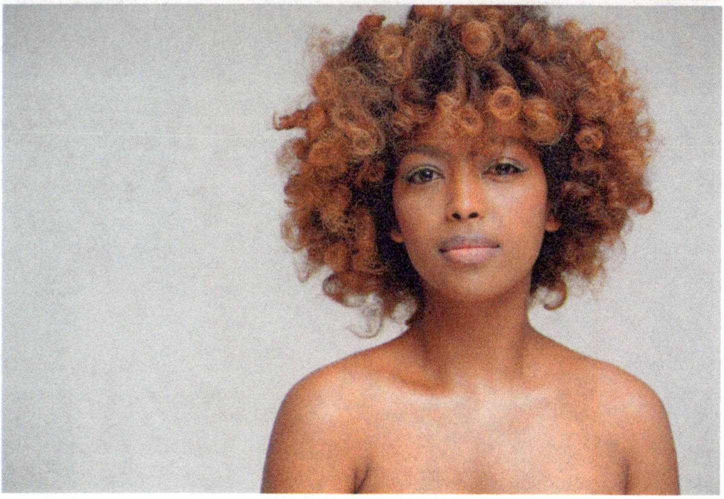

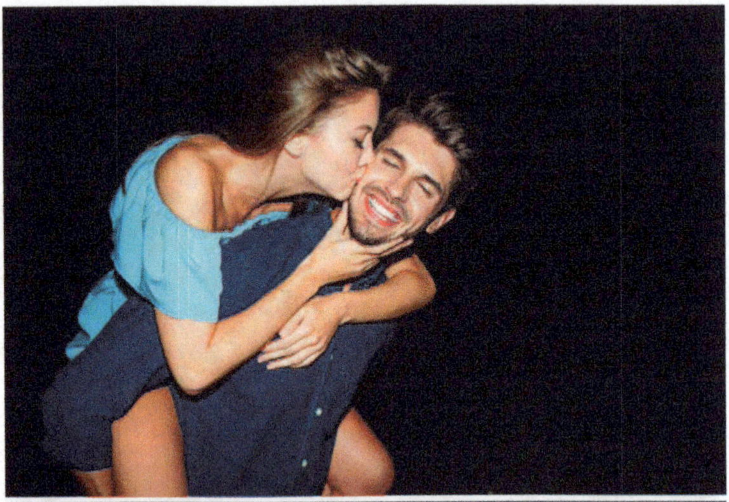

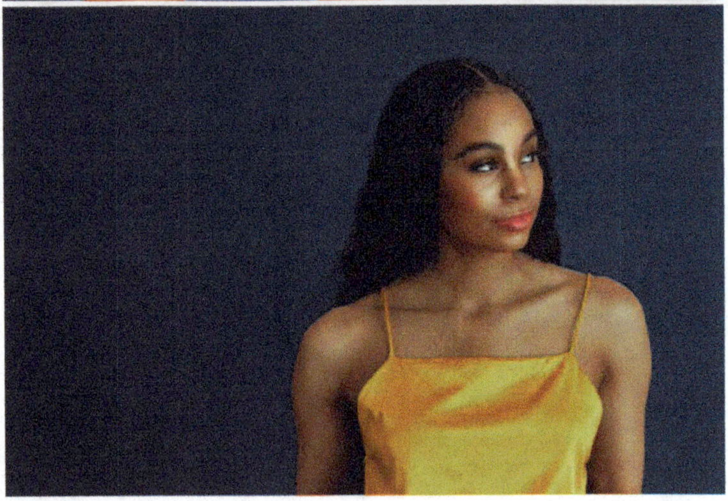

CHIN OVER SHOULDER AI

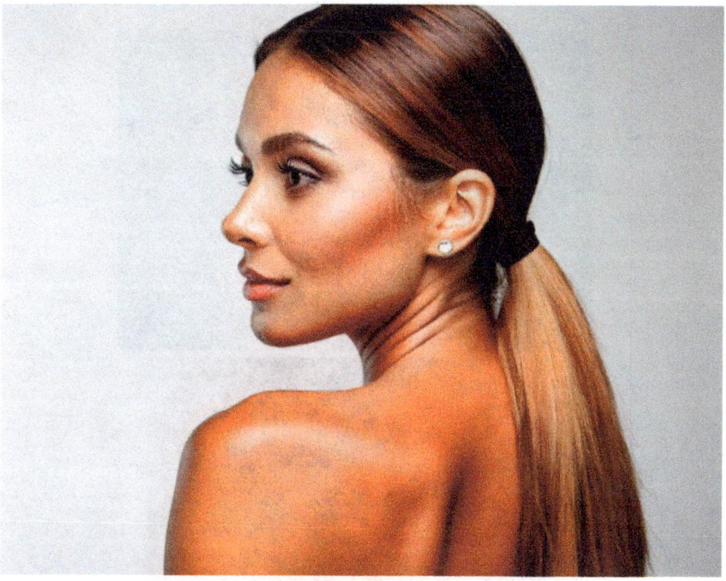

CHIN OVER SHOULDERS STOCK

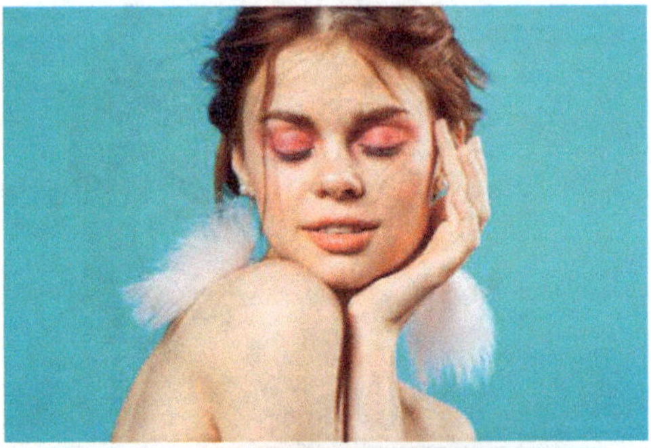
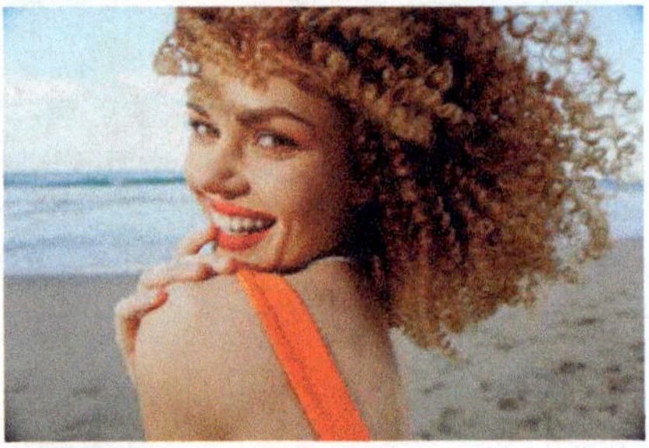
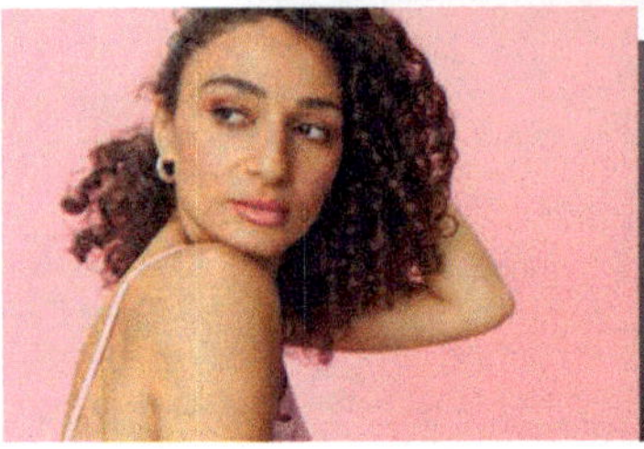
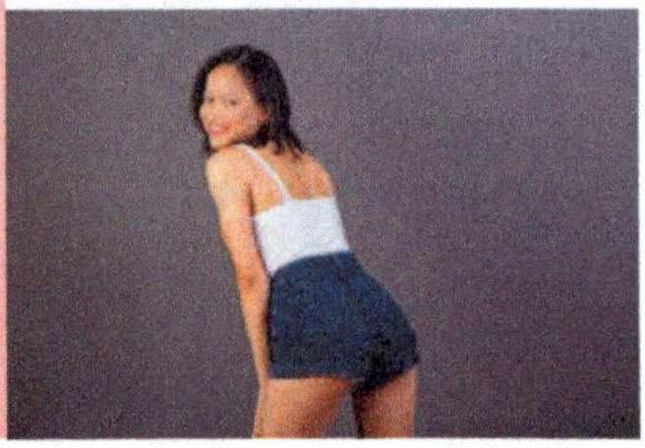

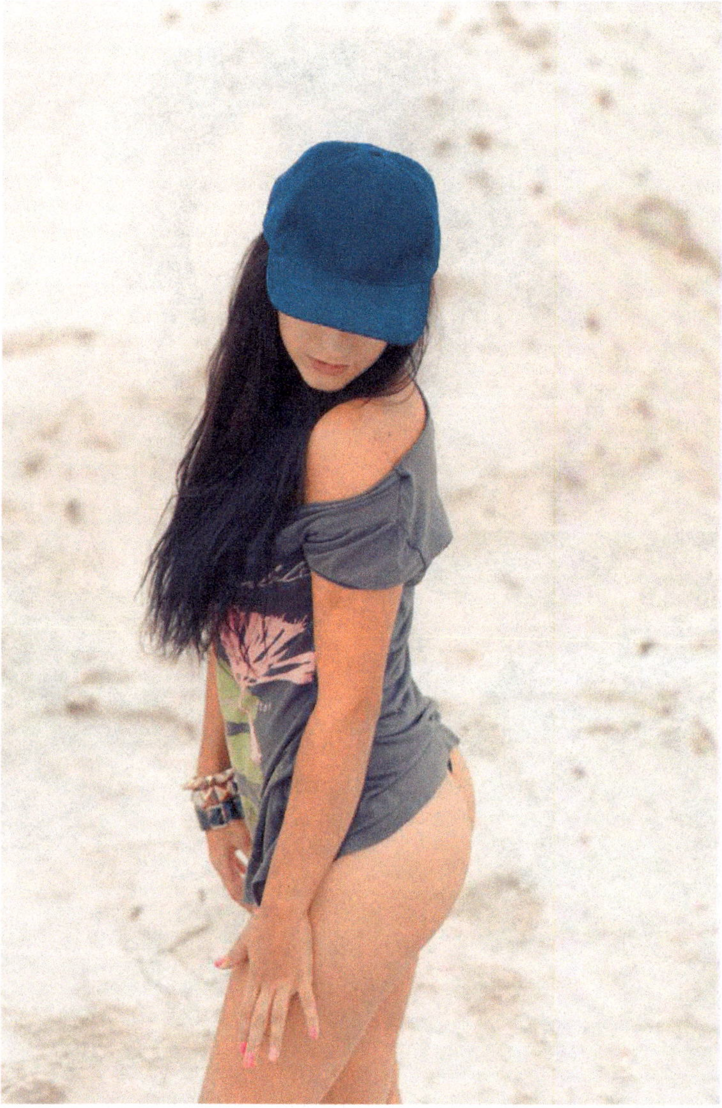

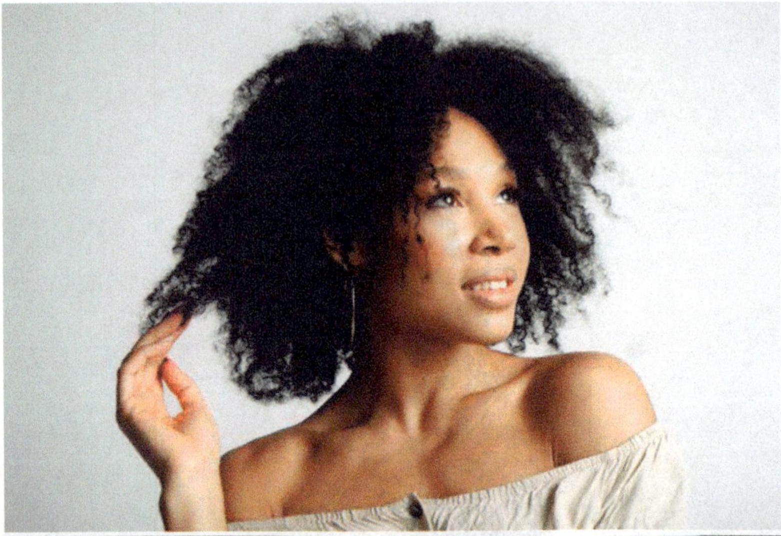

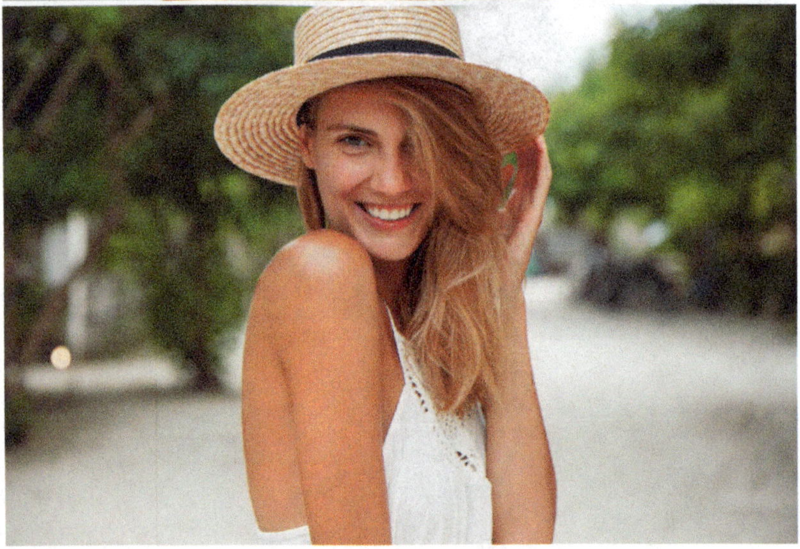

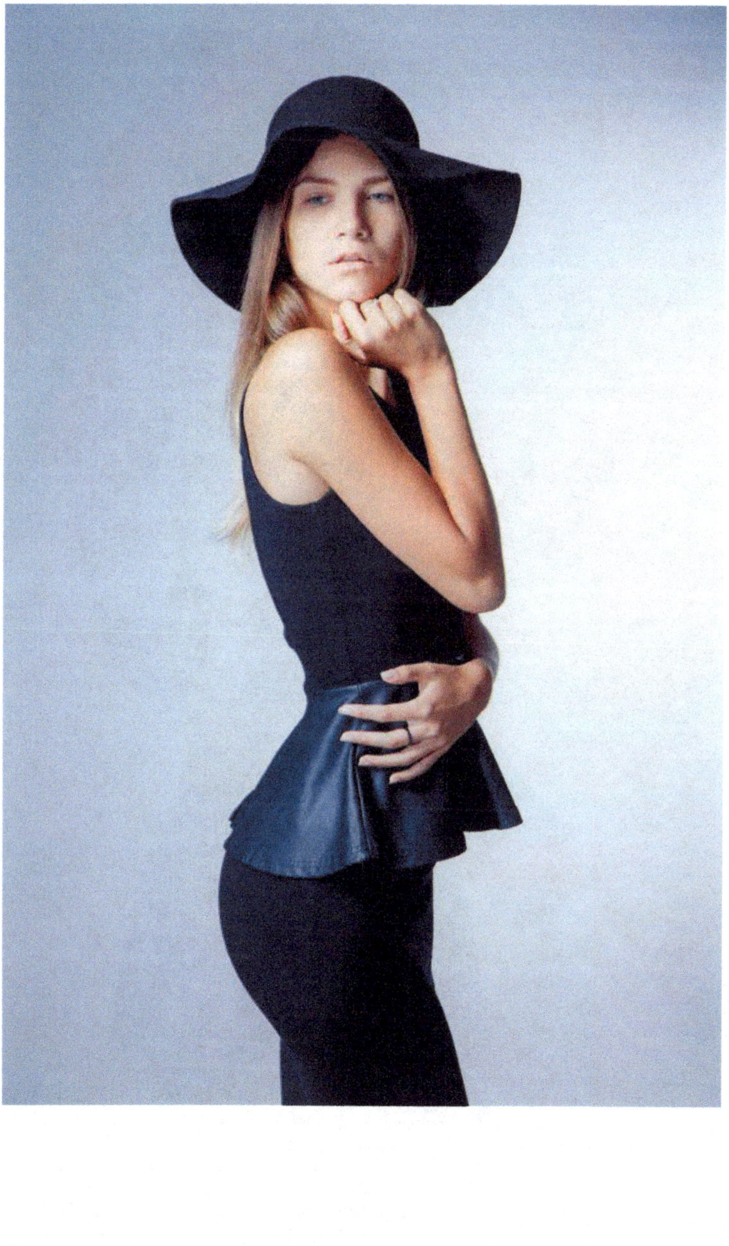

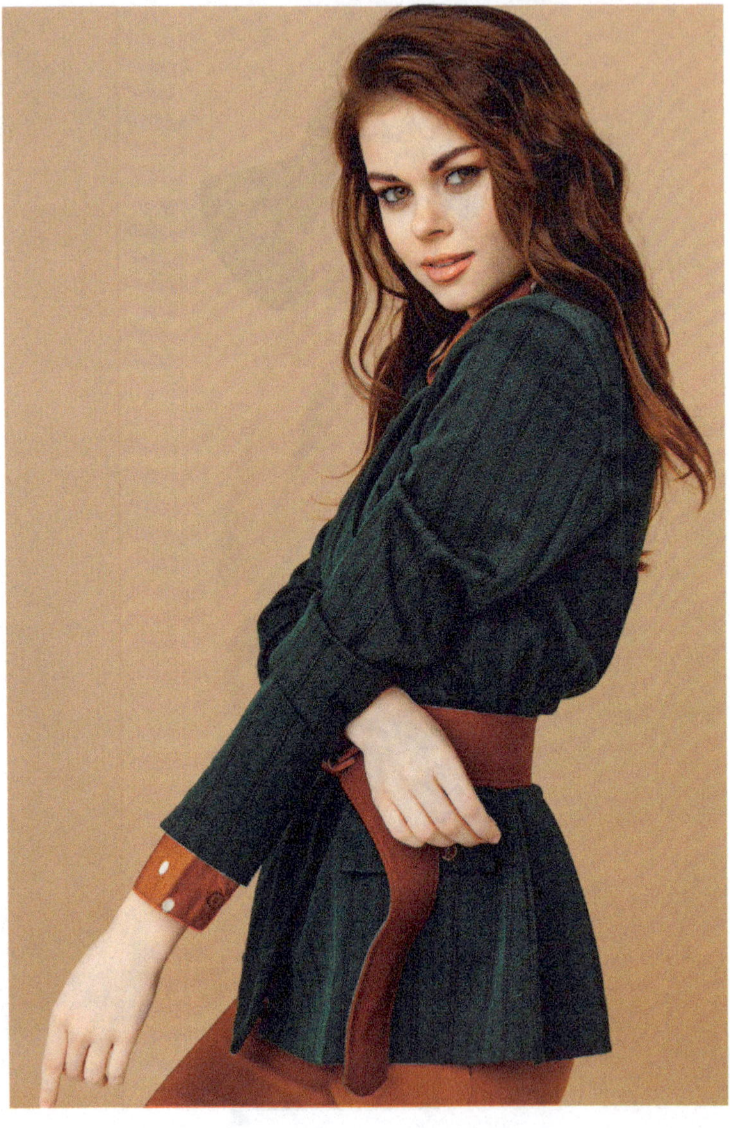

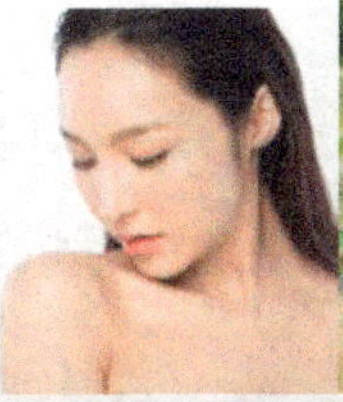
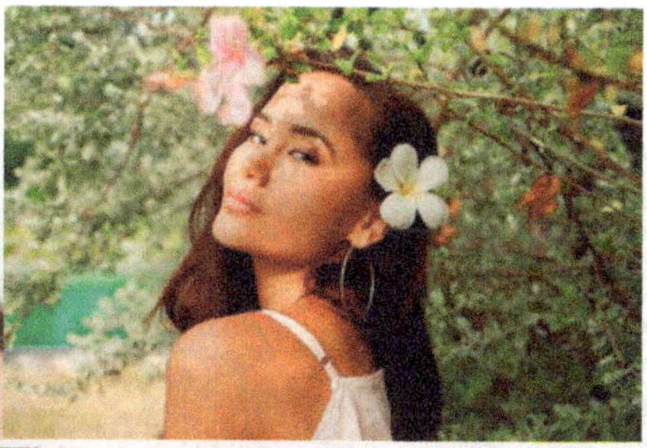
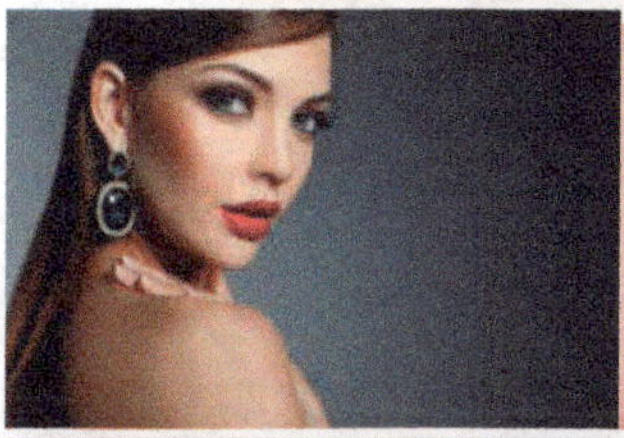
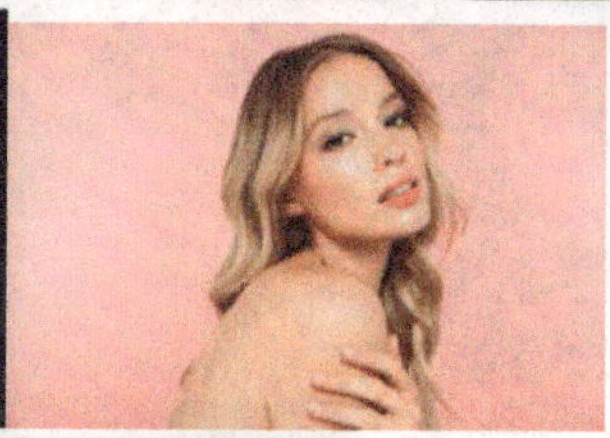
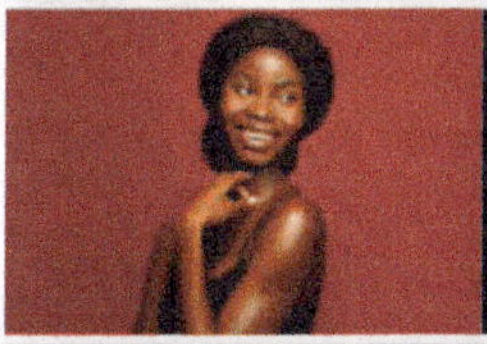
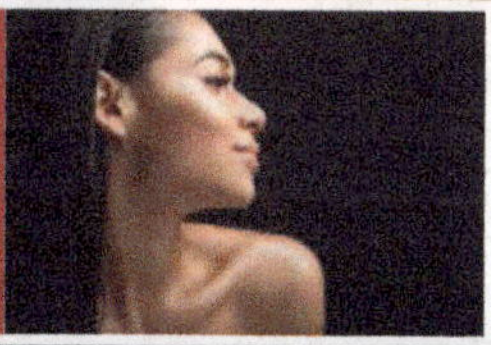
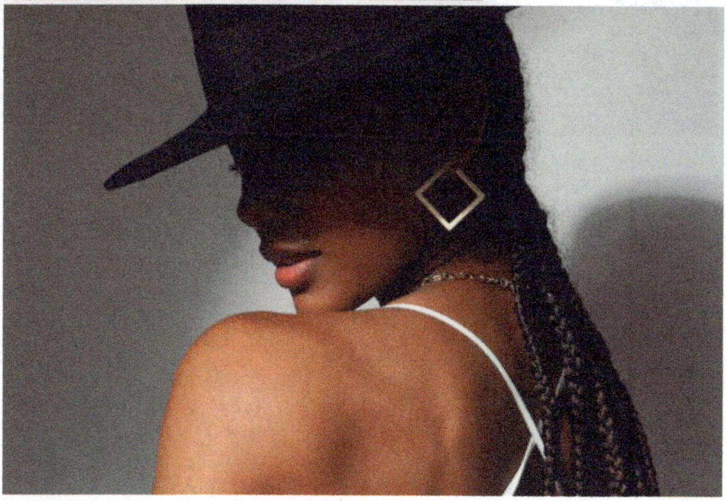

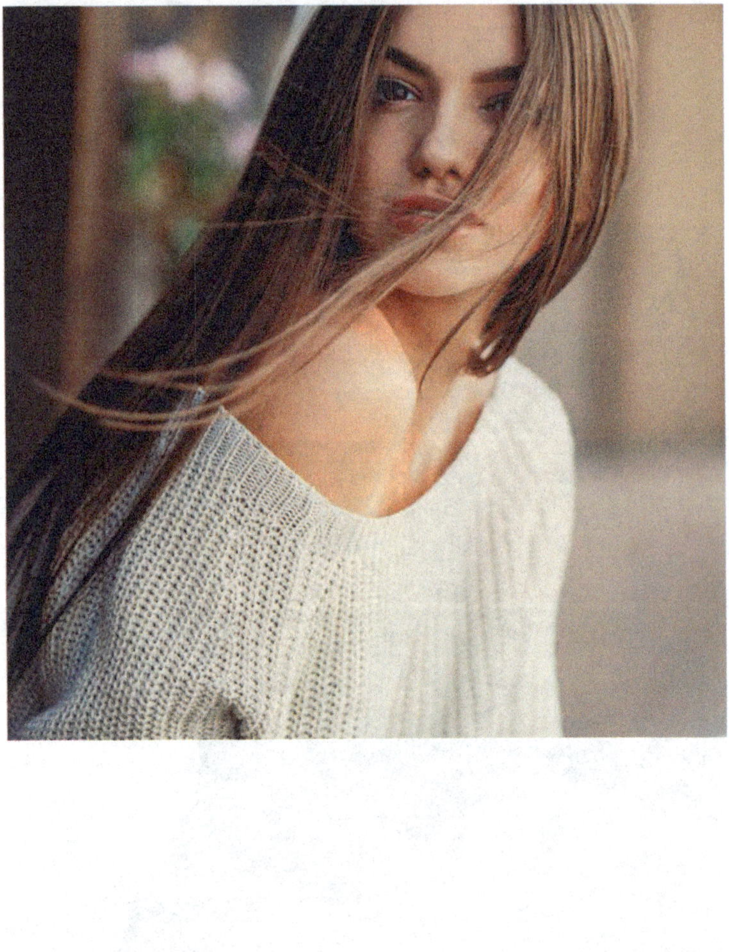

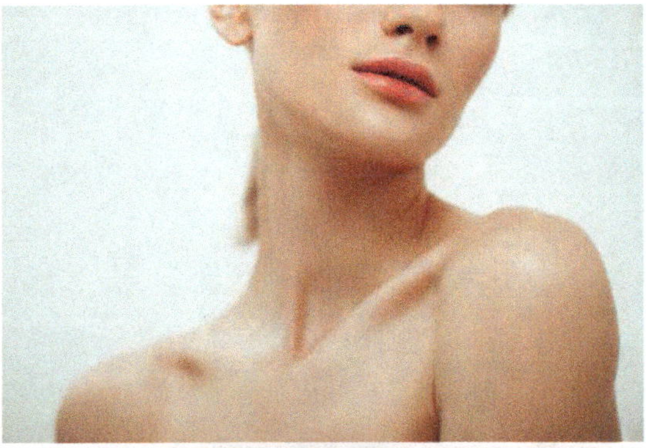

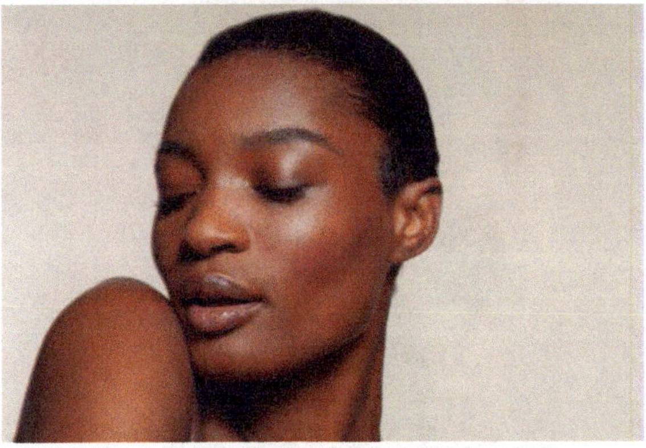

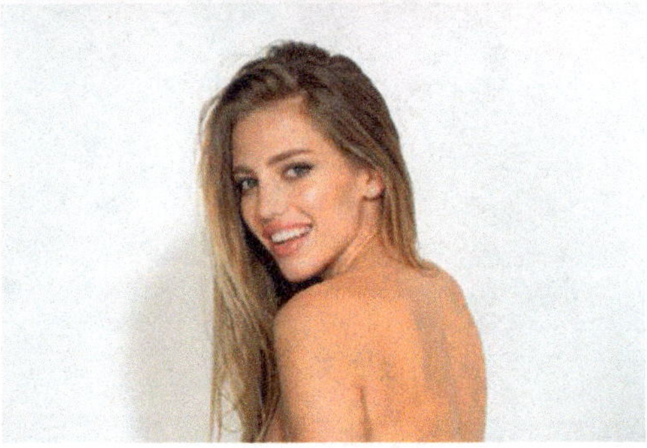

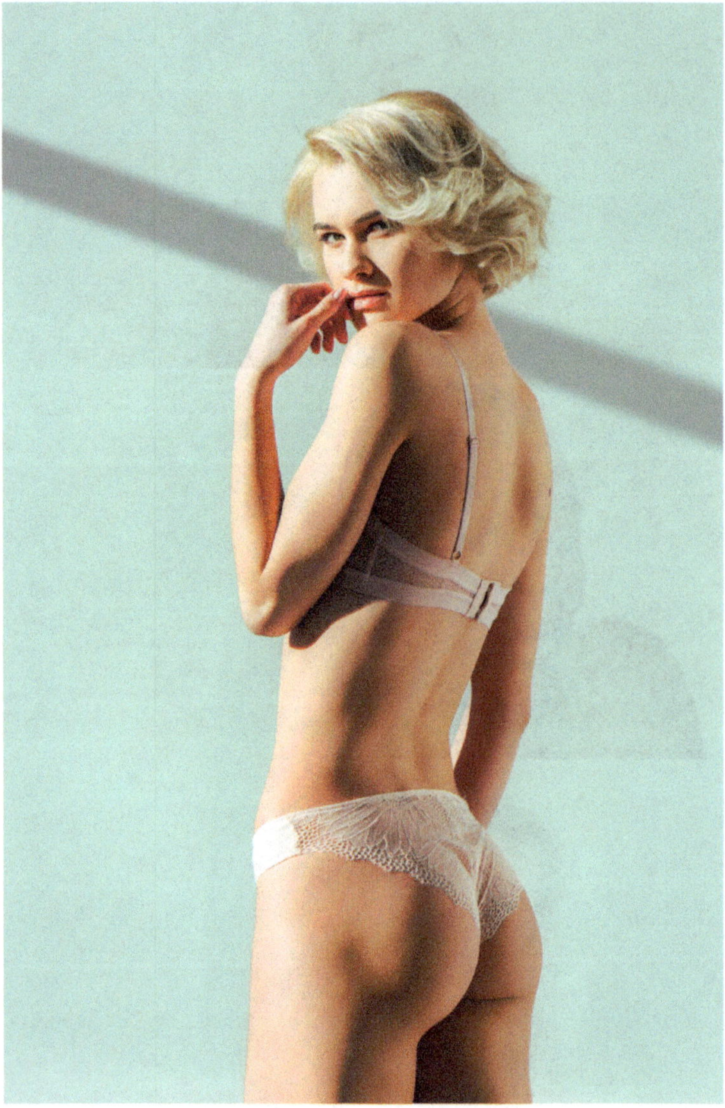

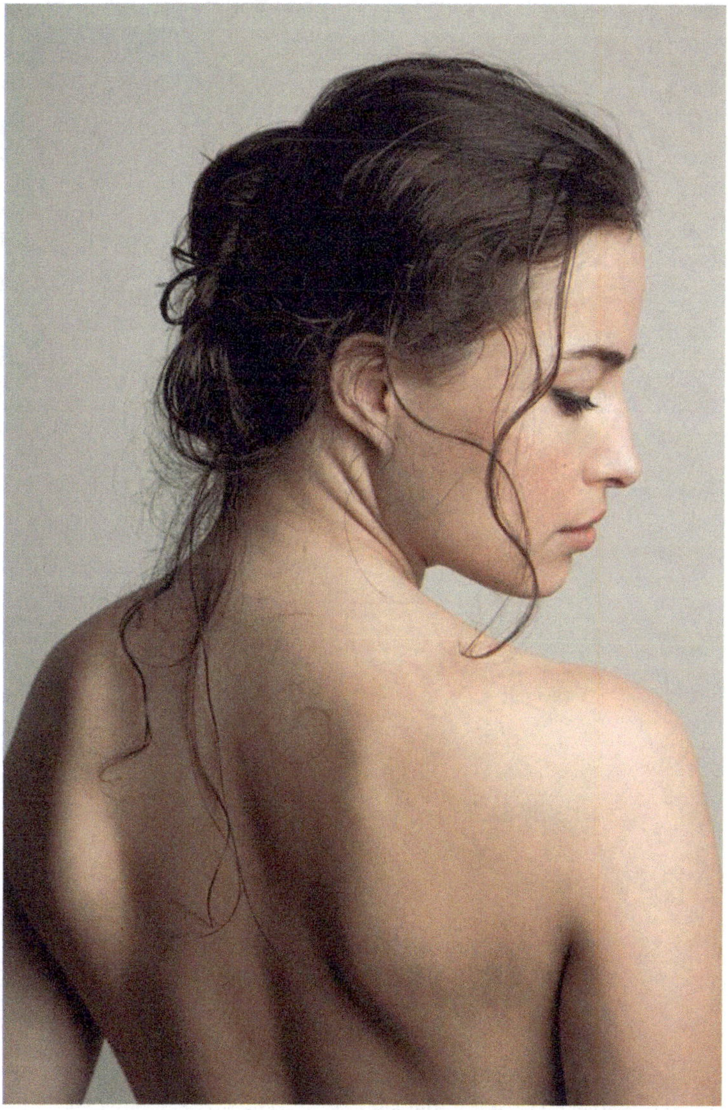

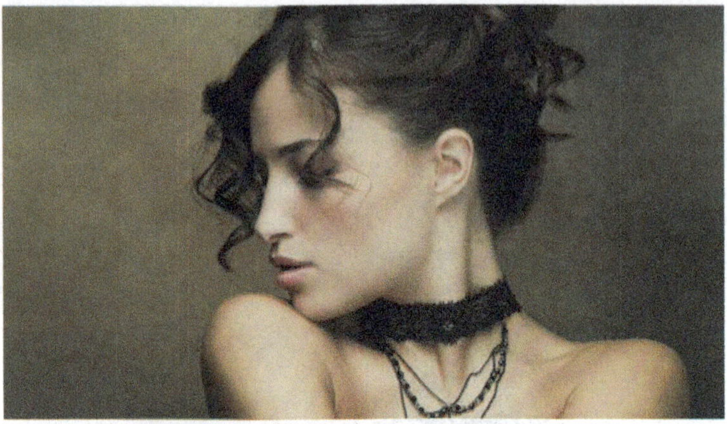

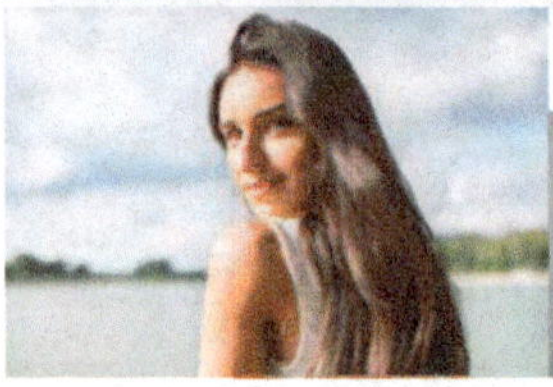

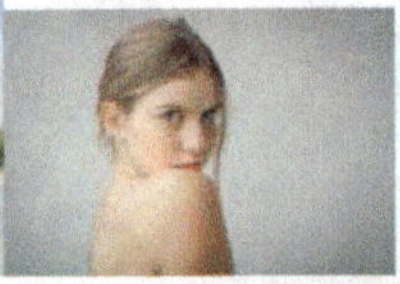

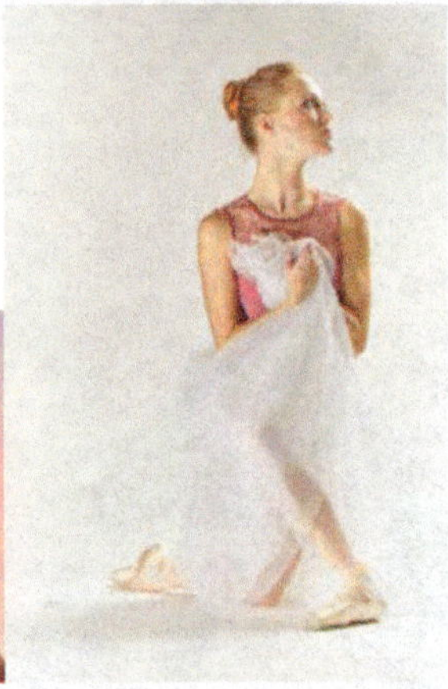

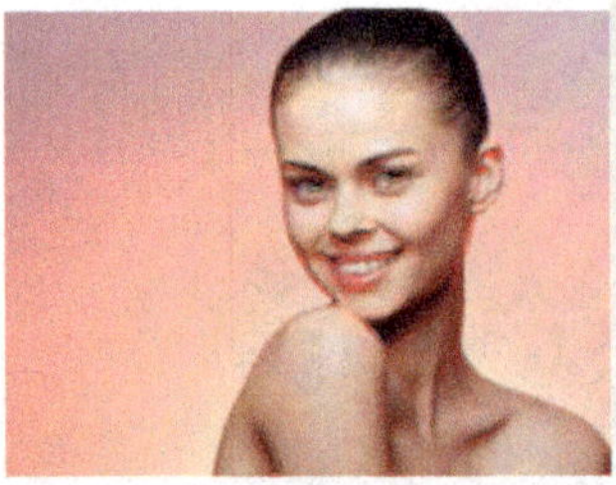

THE FOLLOWING PAGES ARE PAGES FROM THE BOOK "THE FEMALE FORM FROZEN" AND ARE COLLAGES THAT ARE ONE OF A KIND IN ORIGINAL PAPER AND COLOR FOR SALE FOR 1500 DOLLARS A PIECE

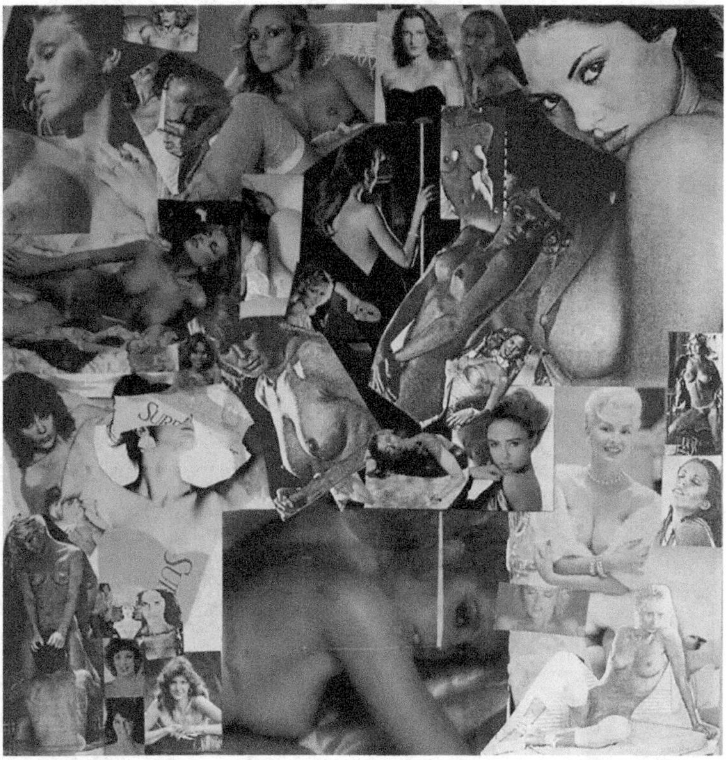

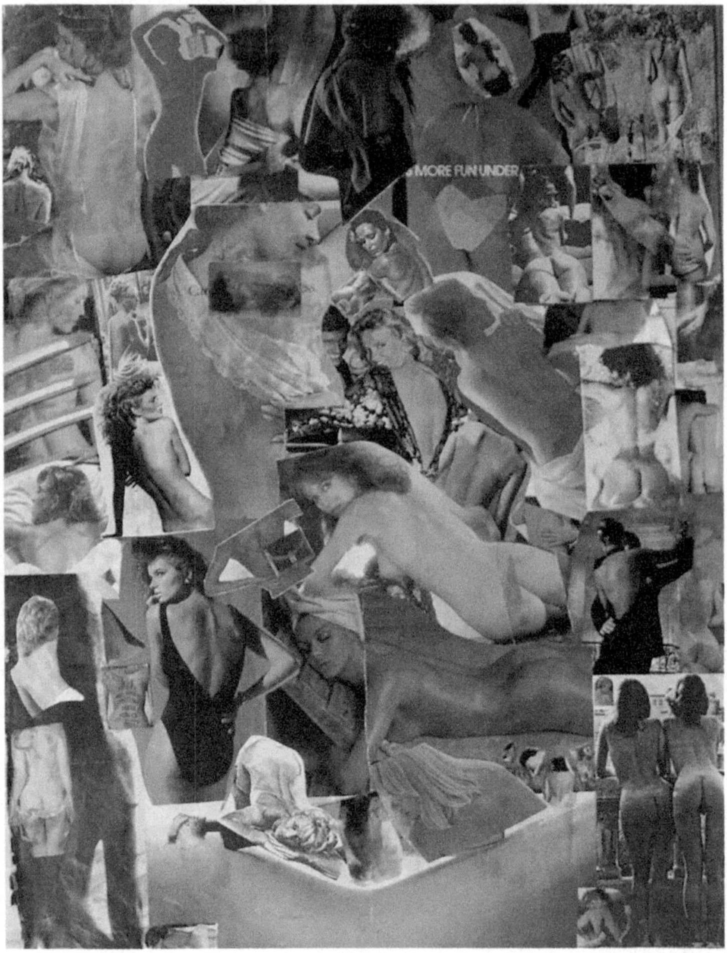
MORE FUN UNDER

ARMS RAISED

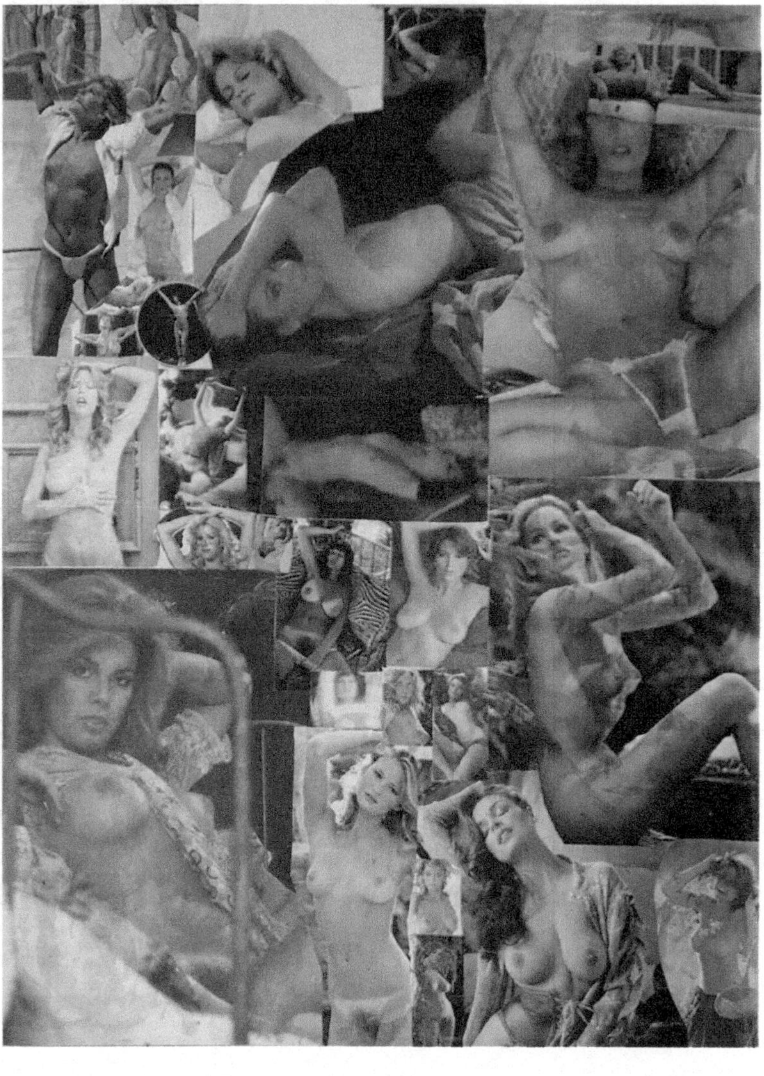

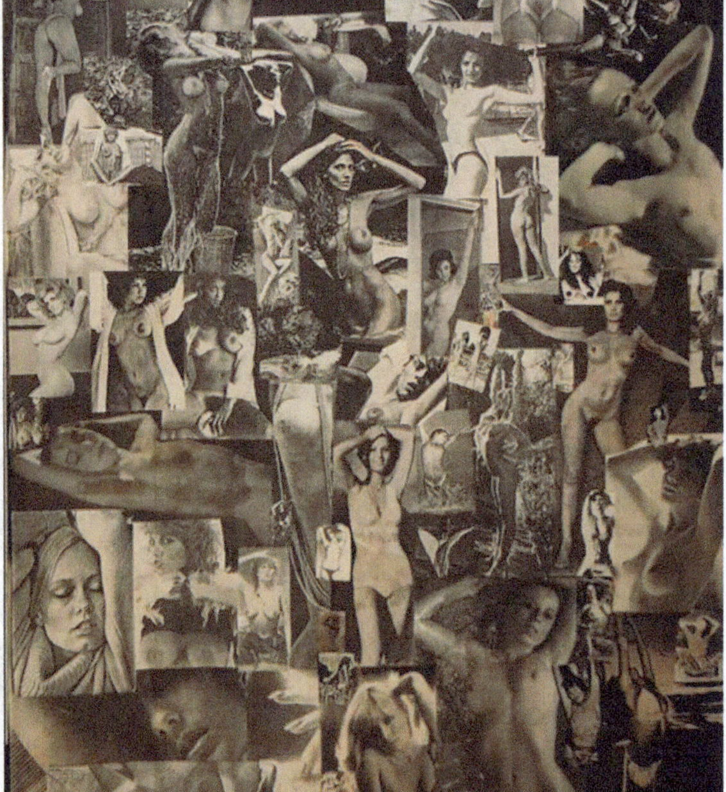

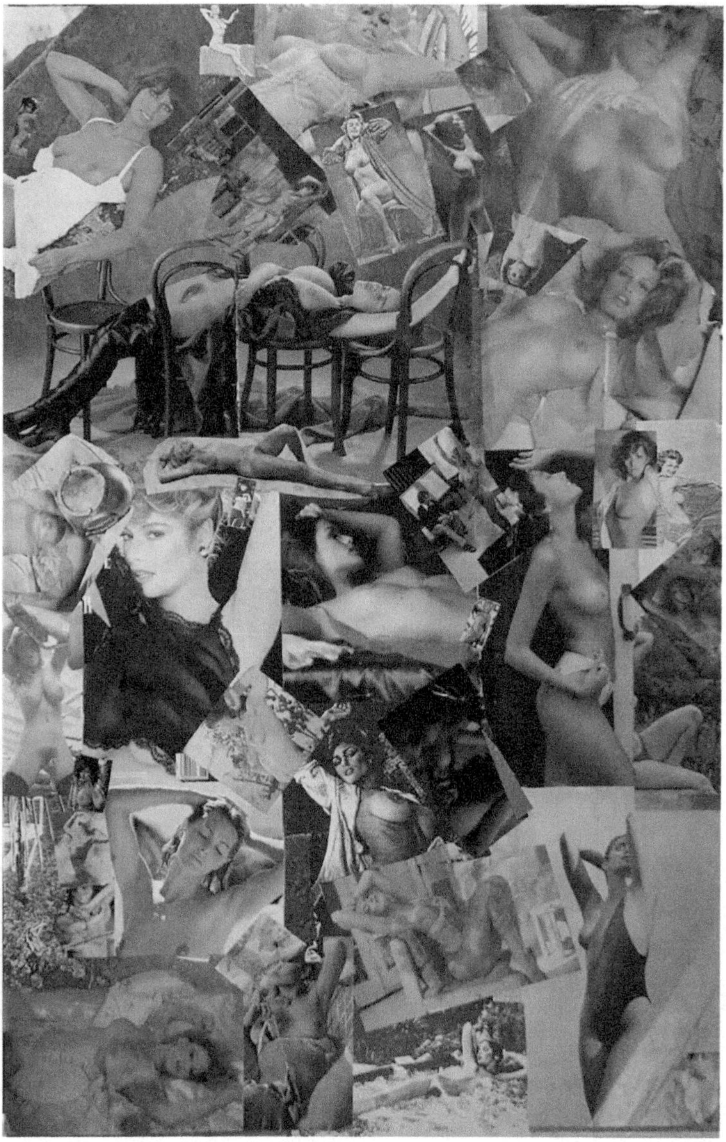

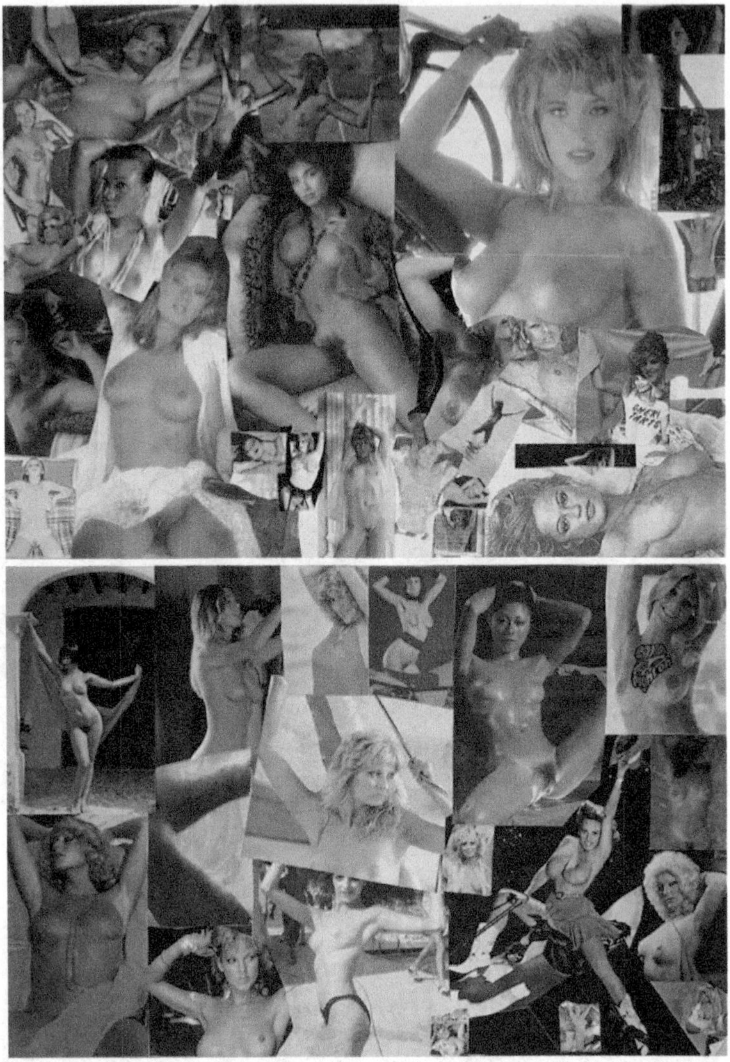

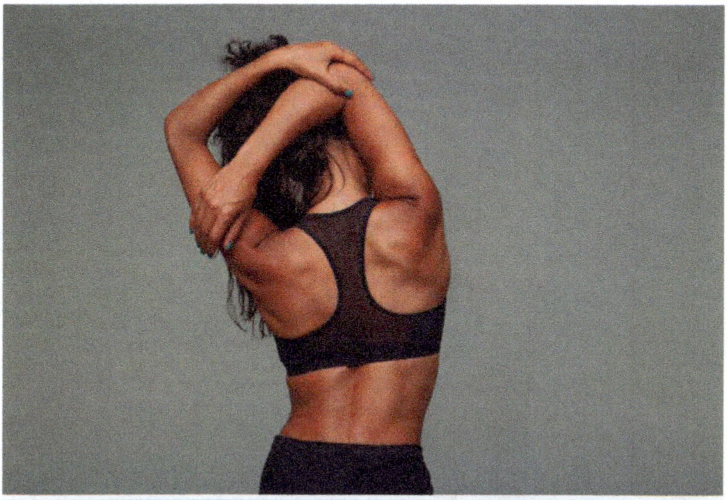

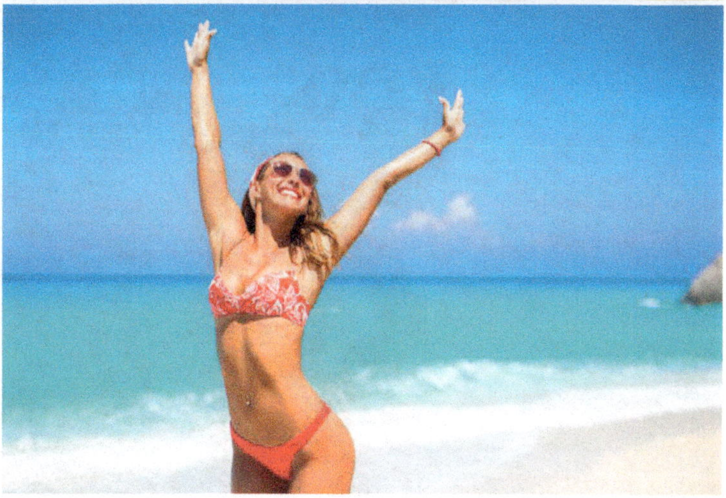

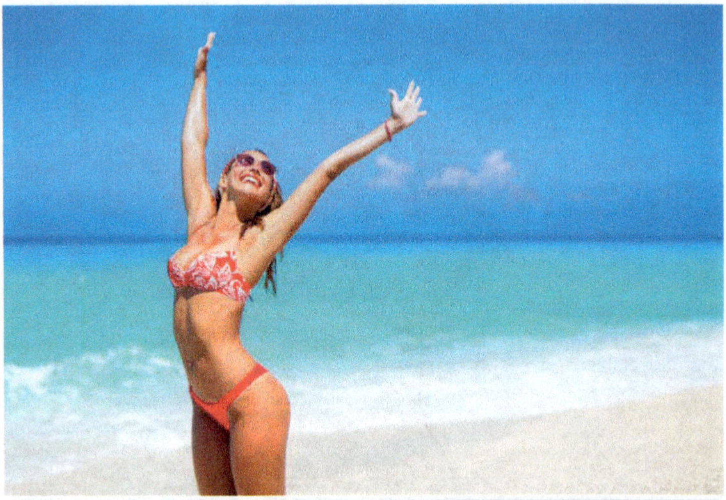

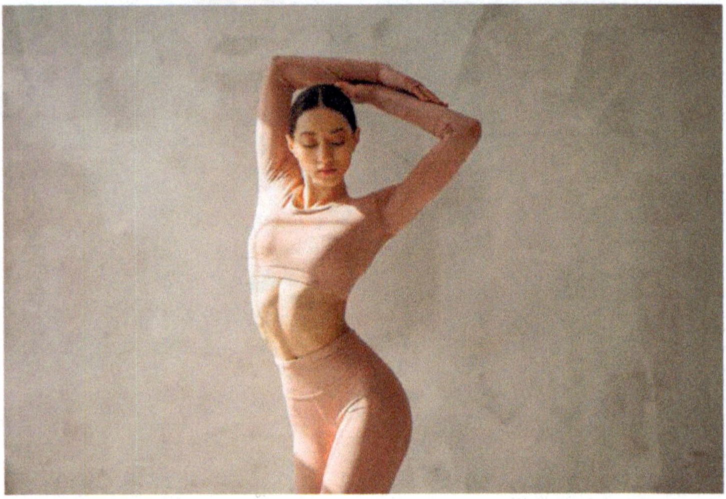

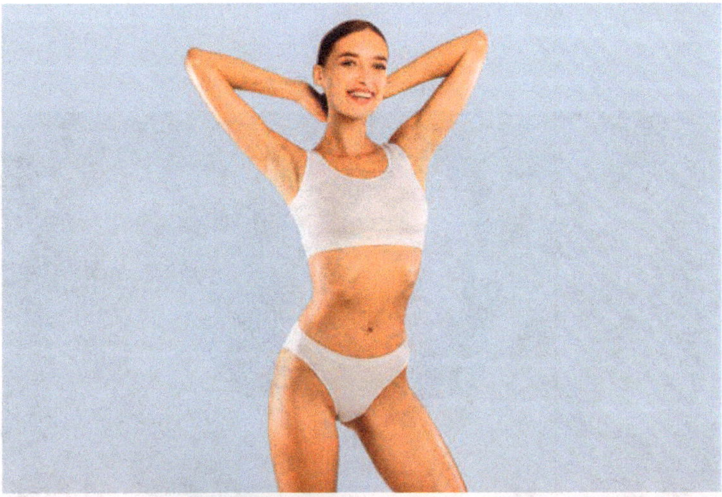

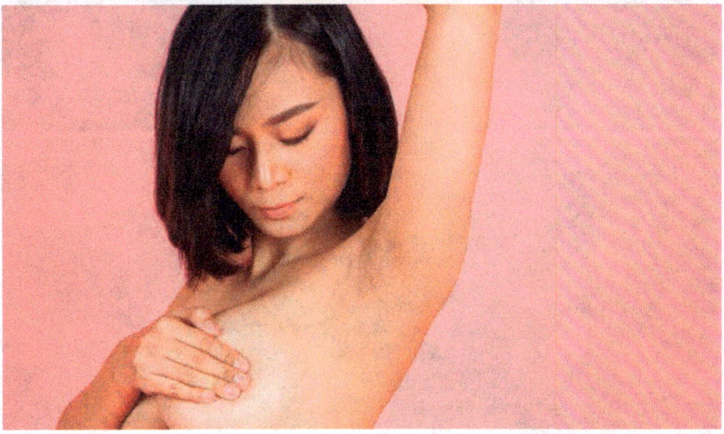

ARMS RAISED AI

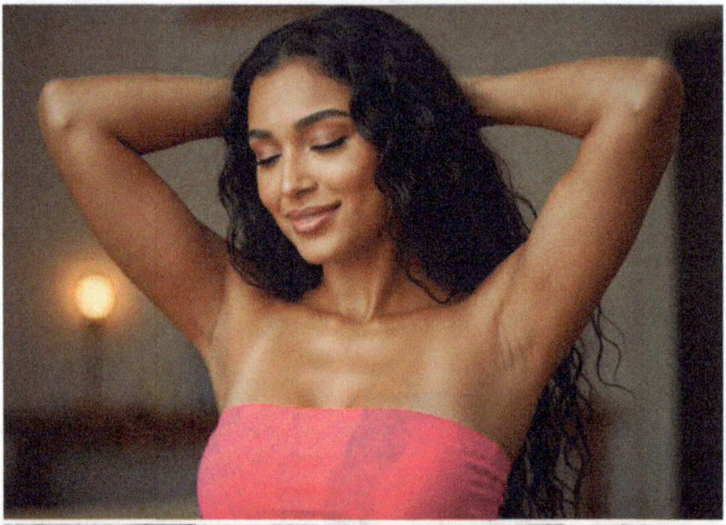

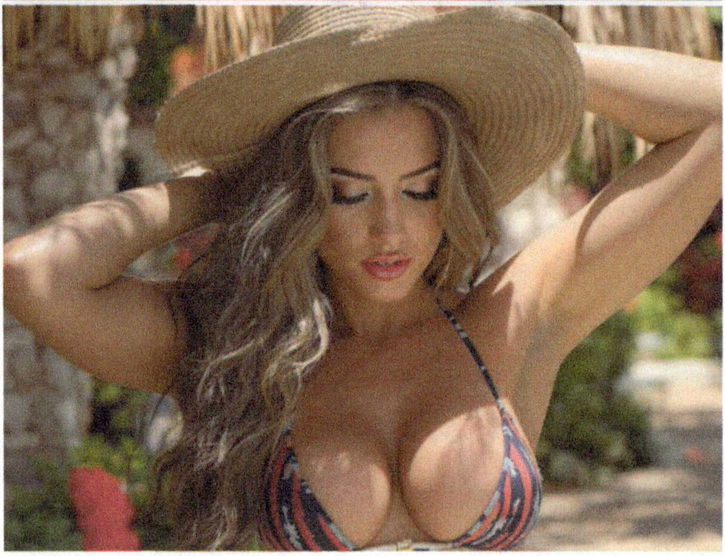

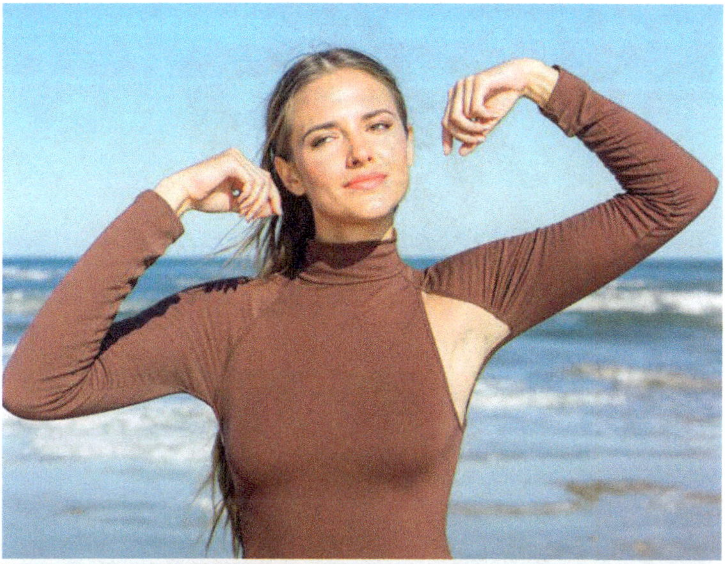

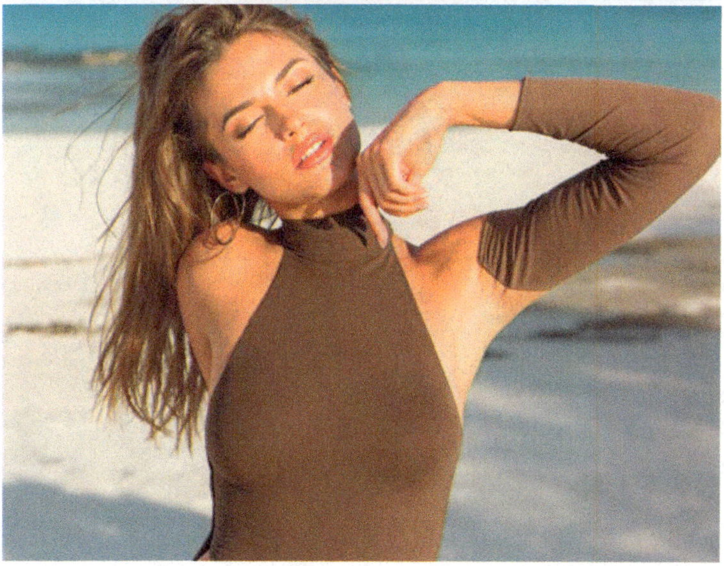

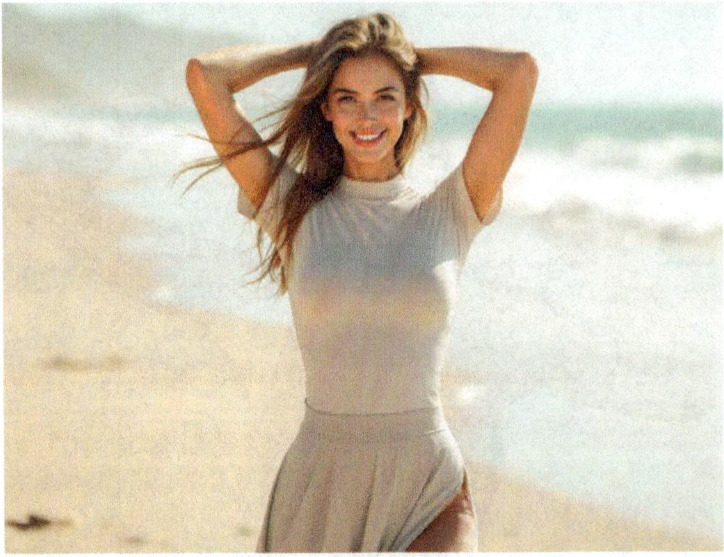

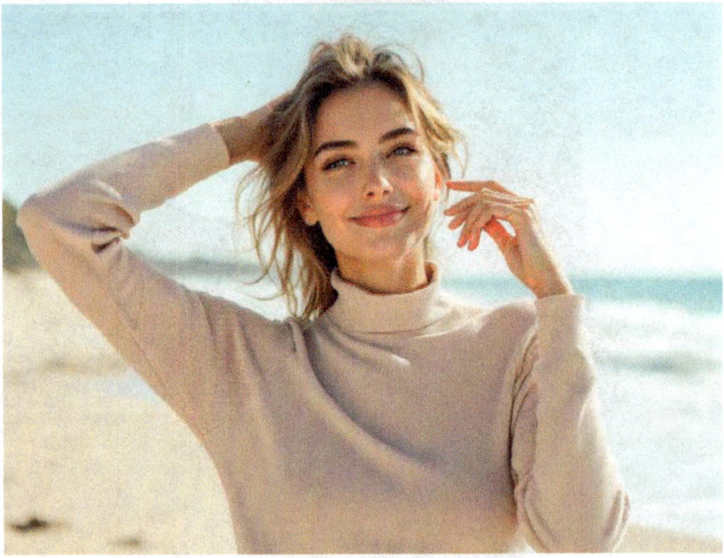

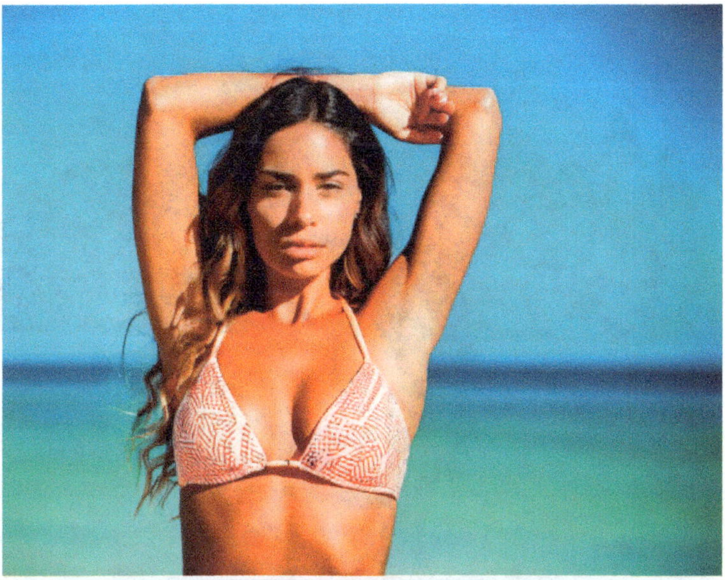

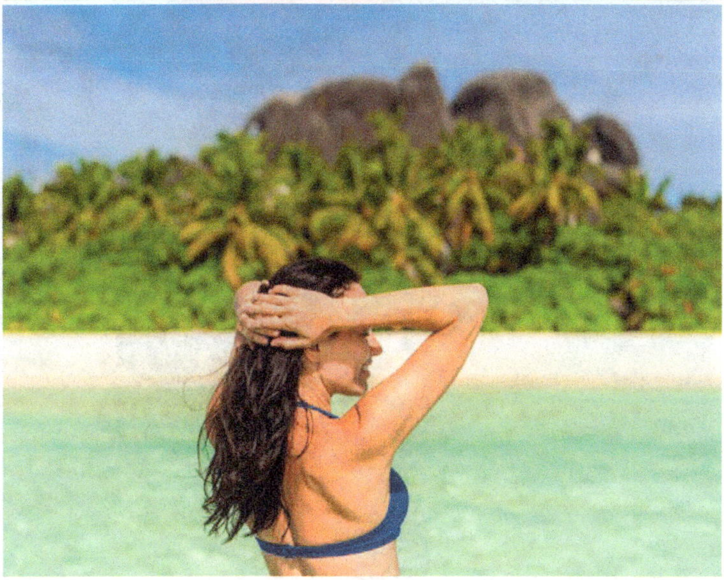

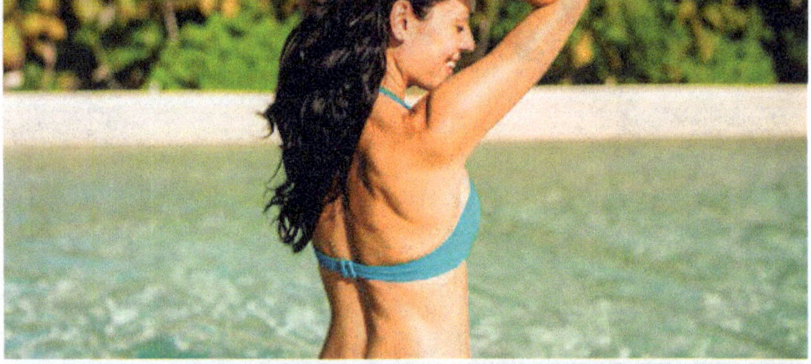

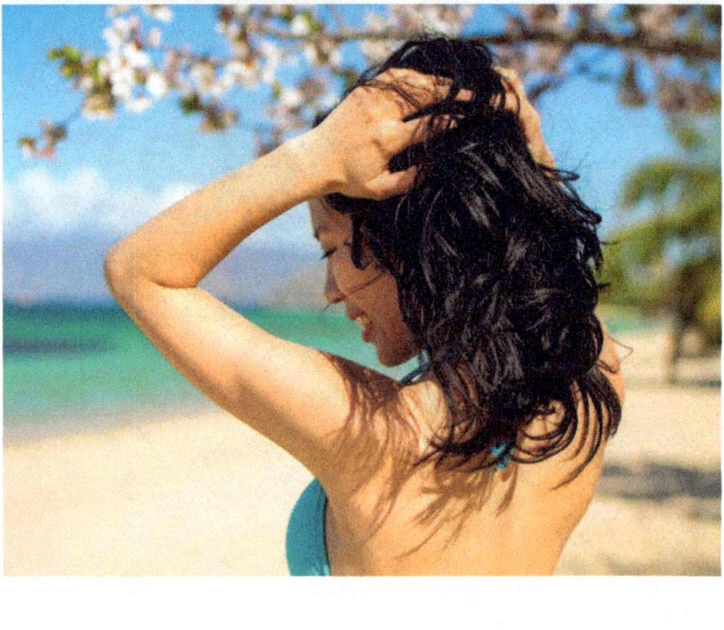

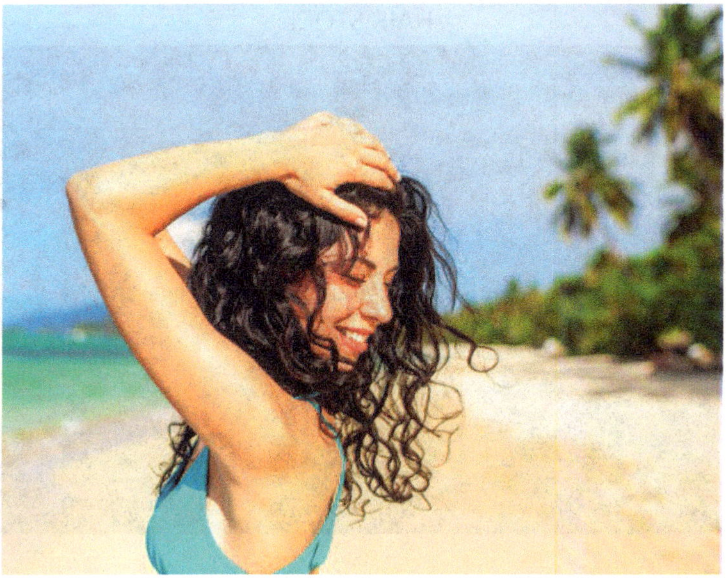

HAIR STOCK

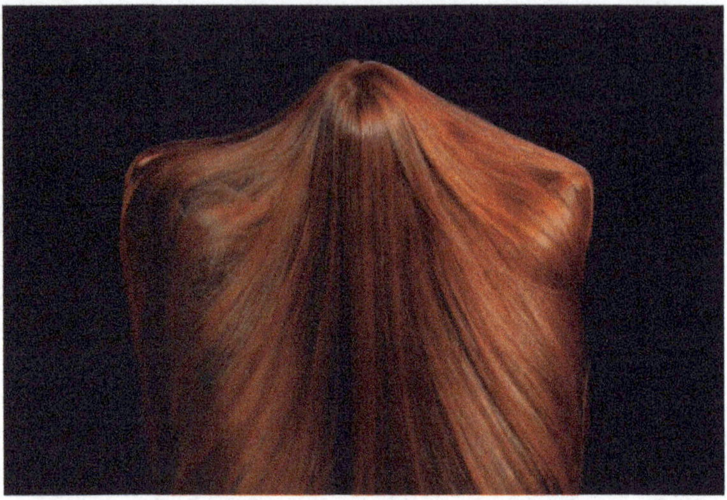

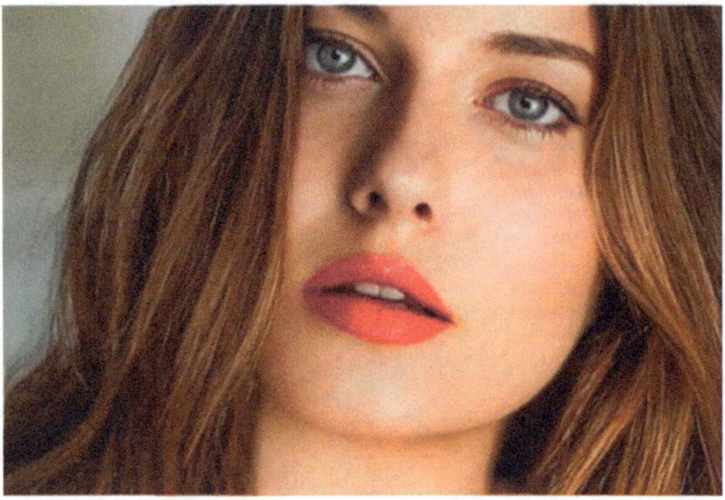

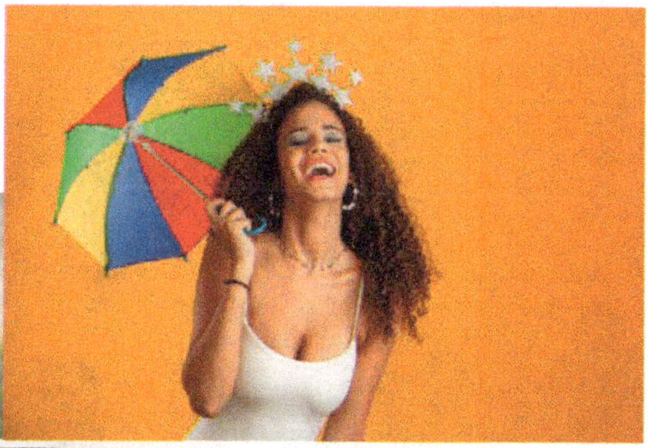
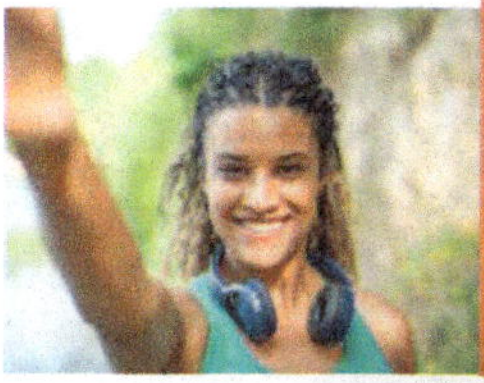
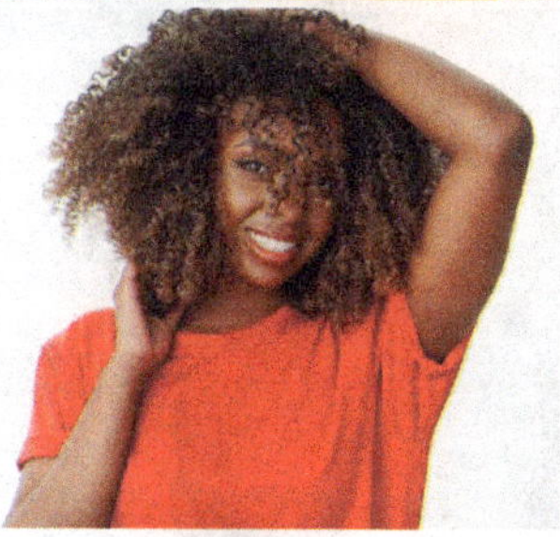
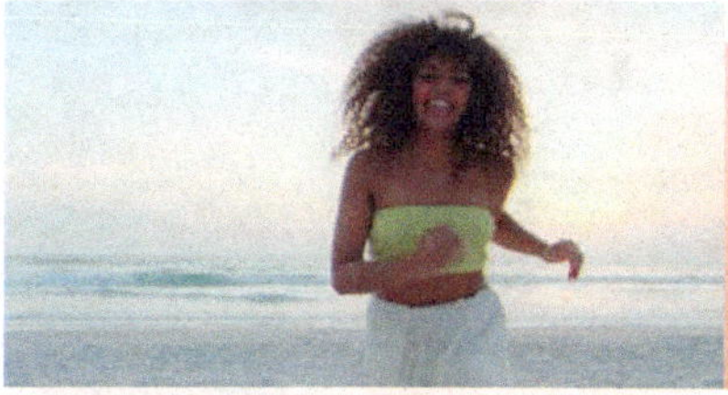
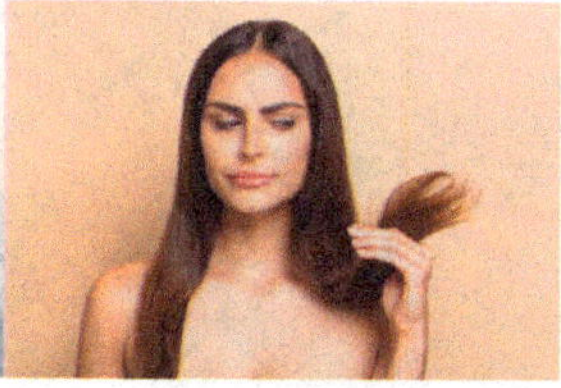
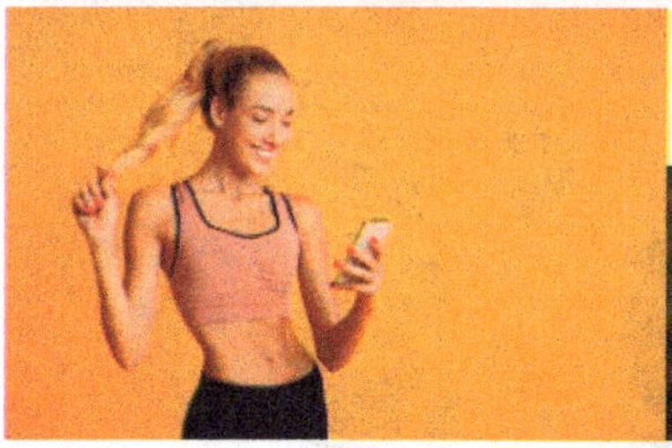
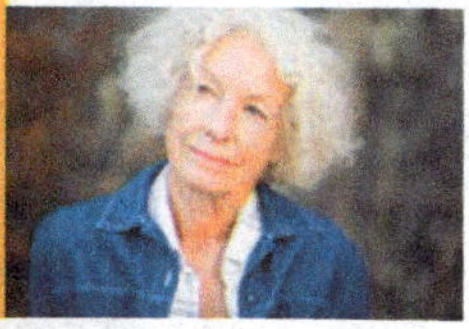

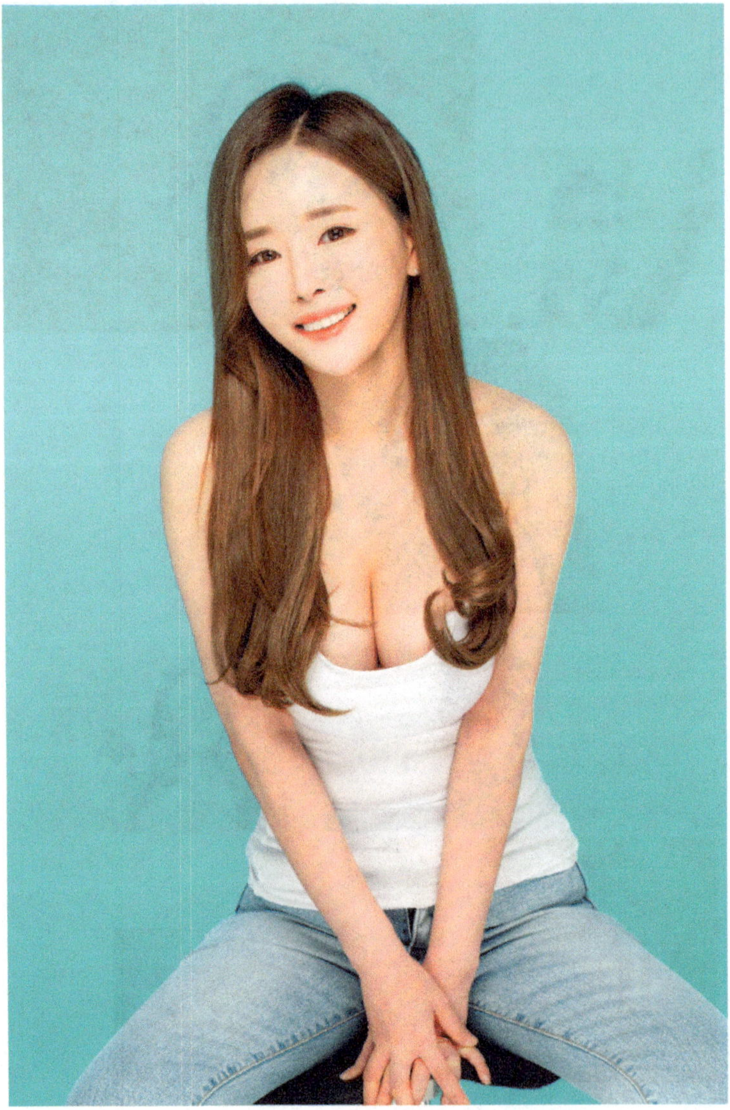

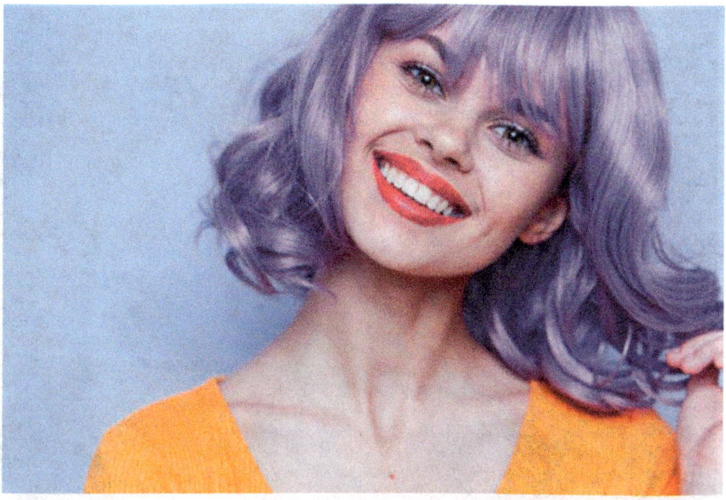

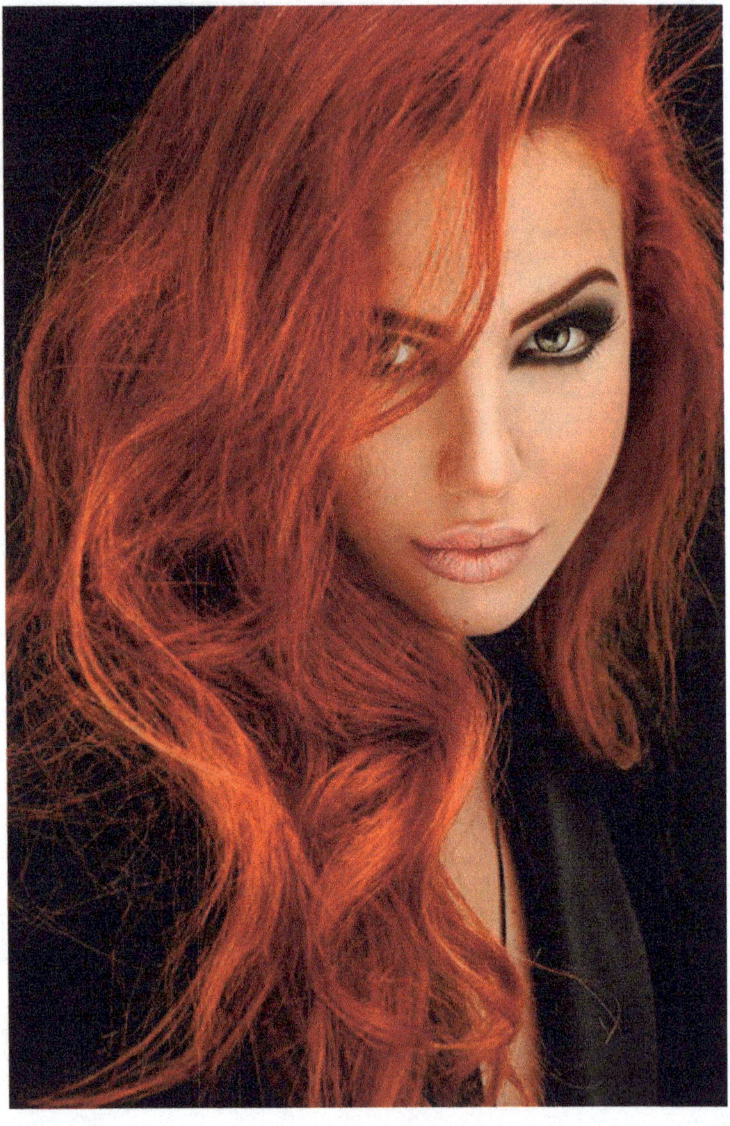

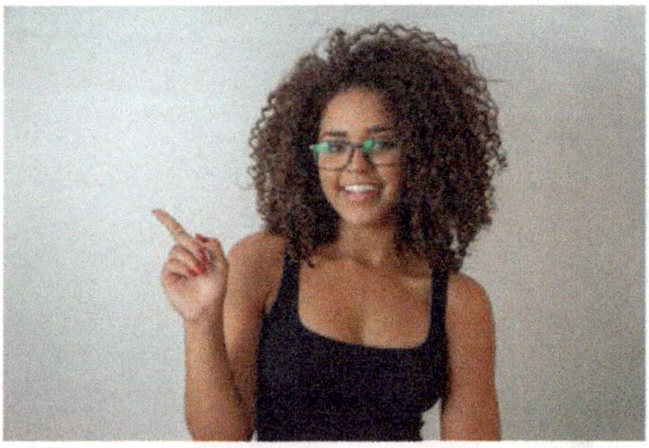
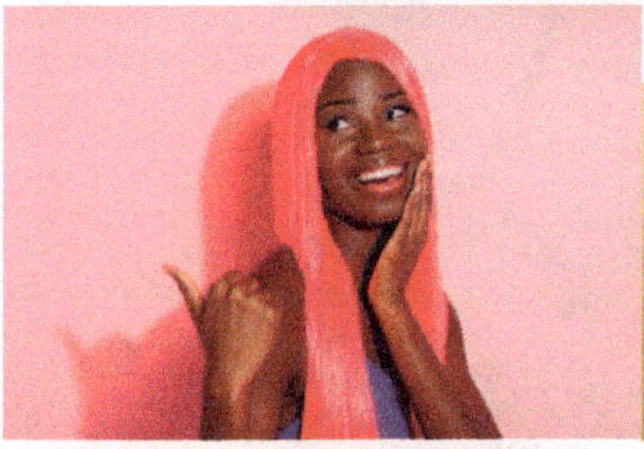
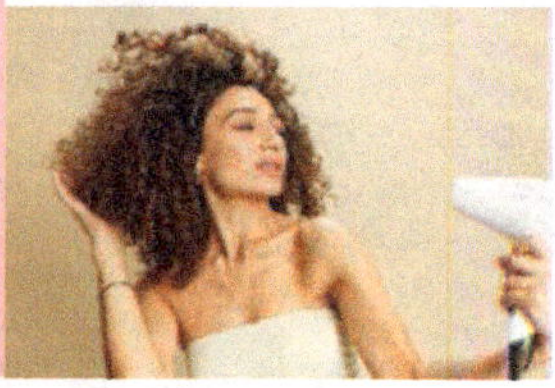
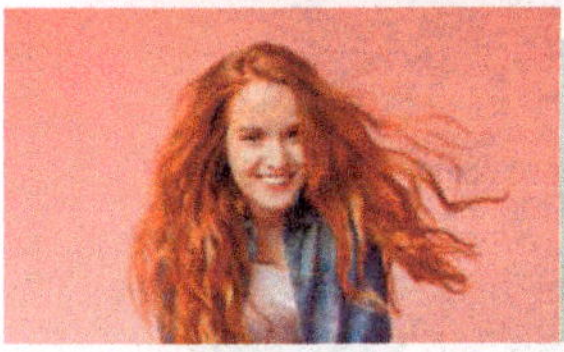
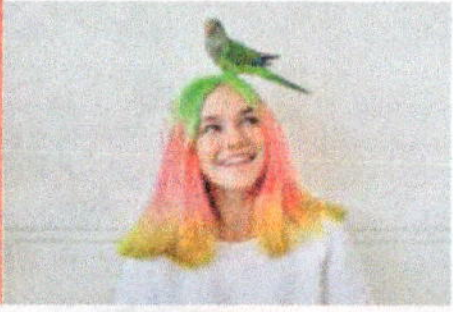
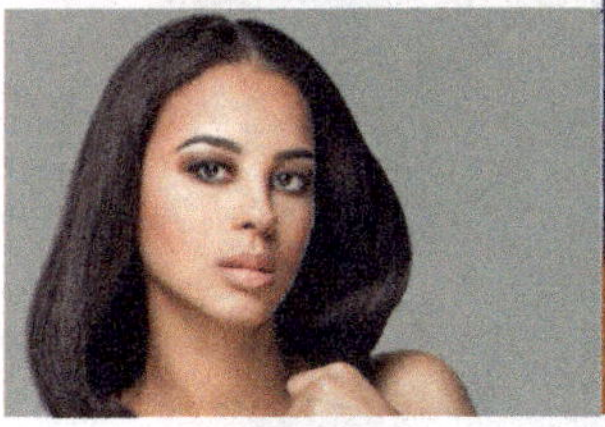
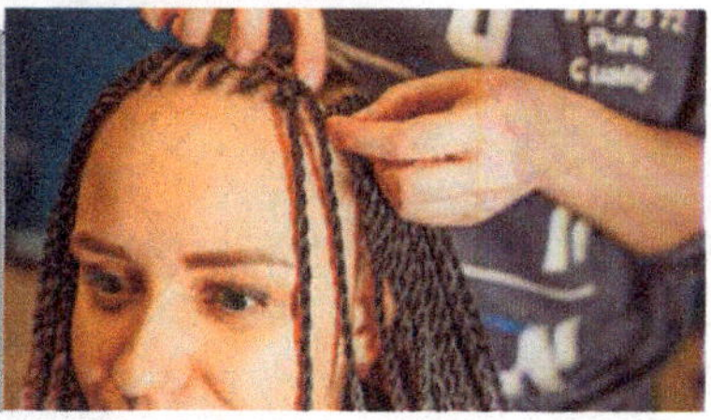
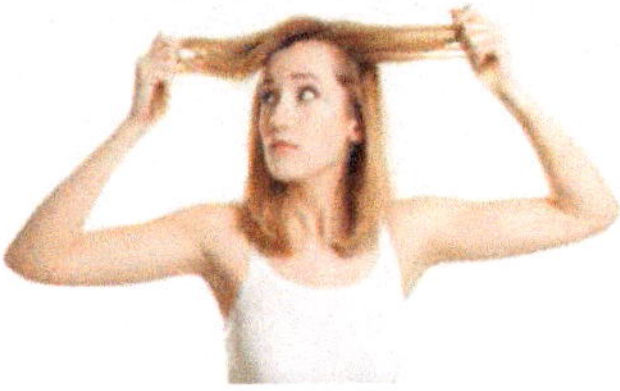
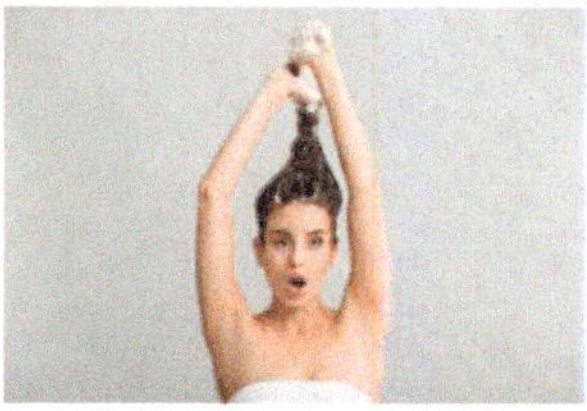

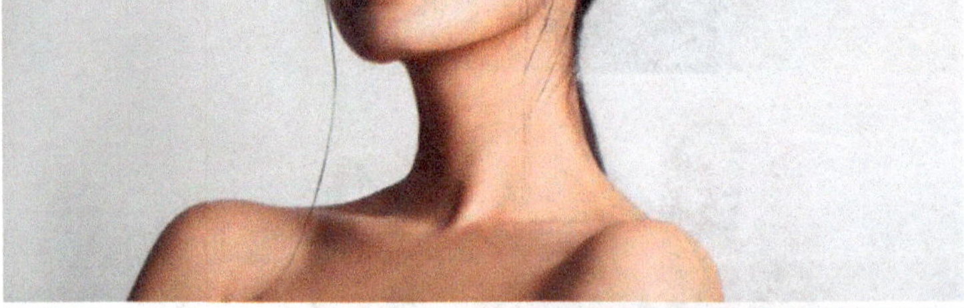

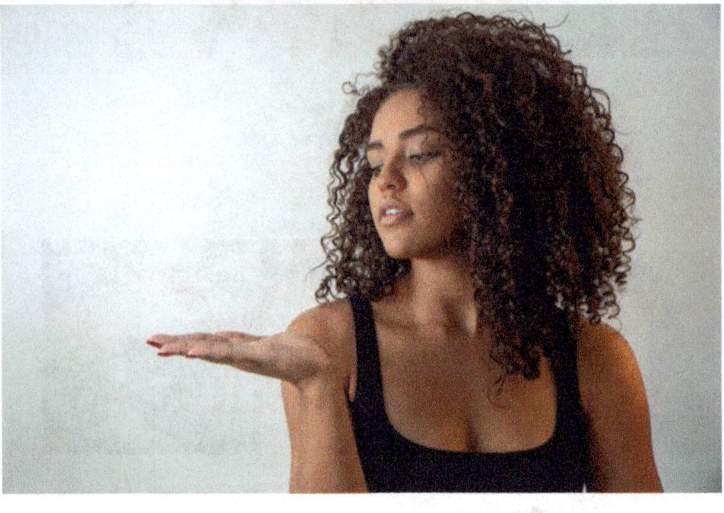

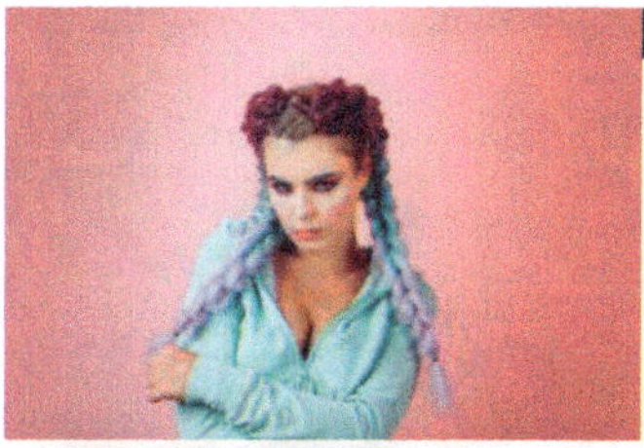

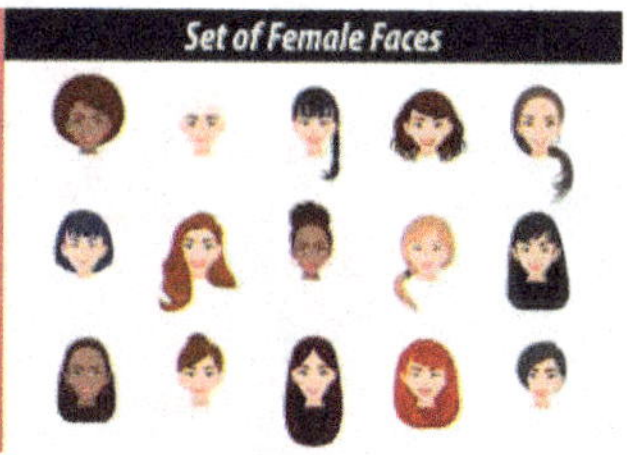
Set of Female Faces

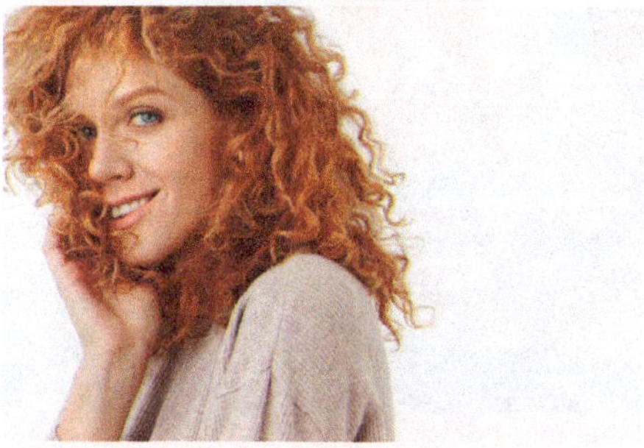

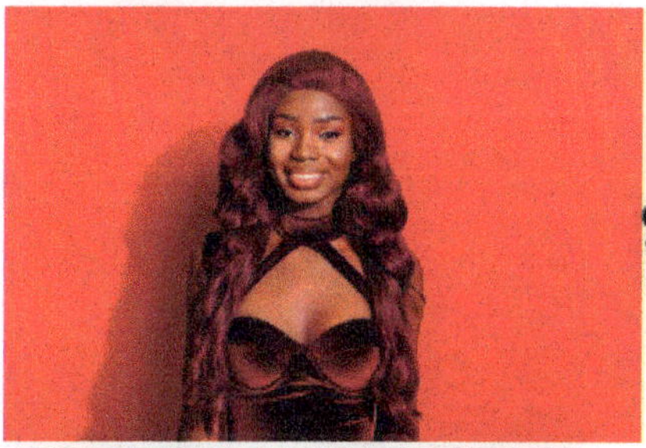

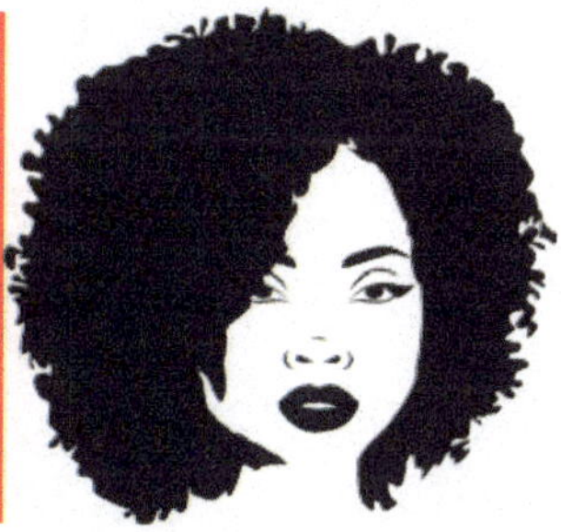

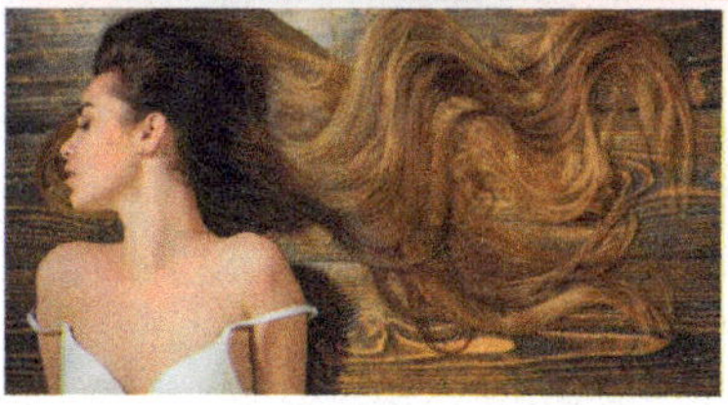

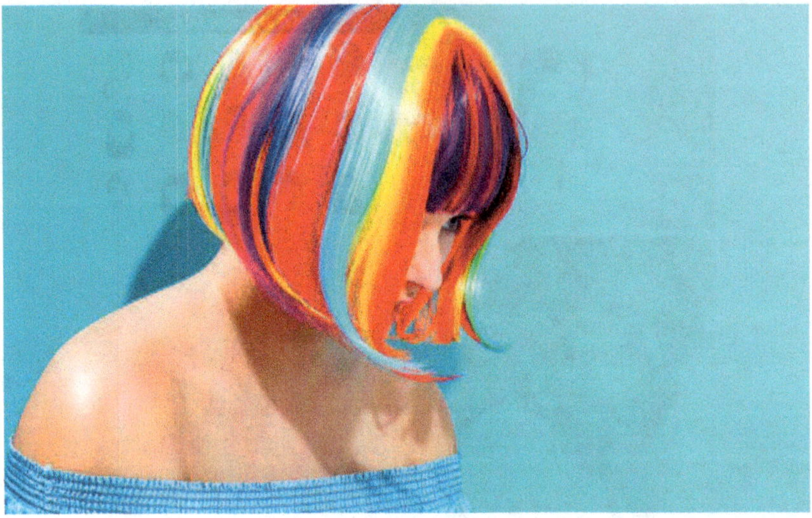

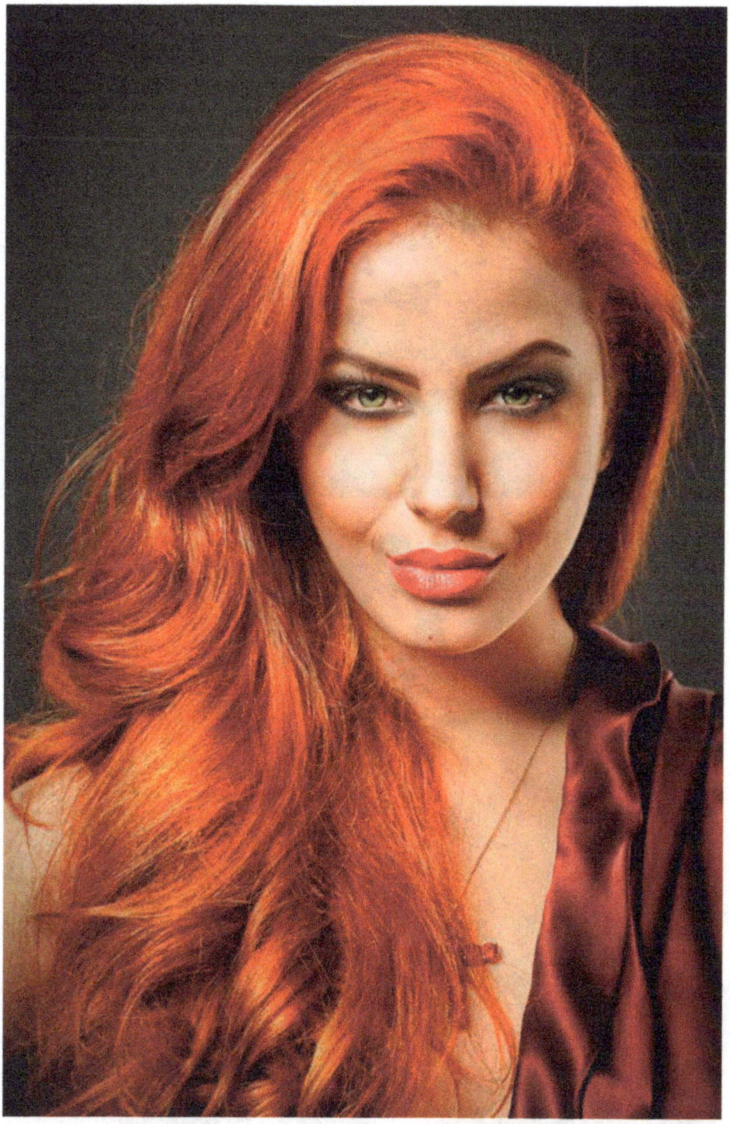

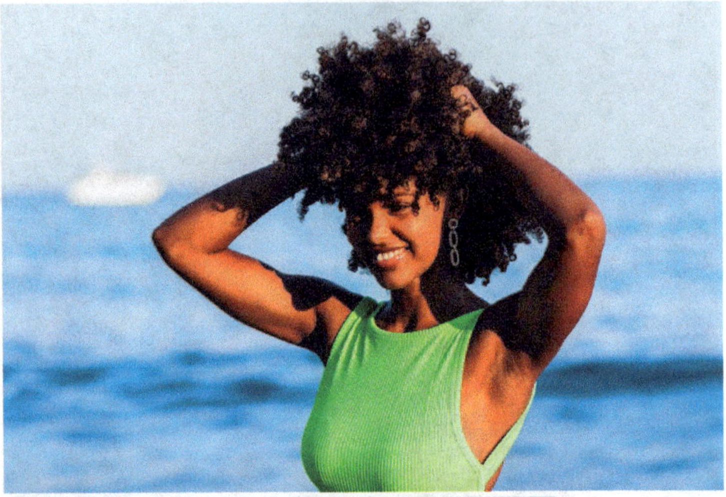

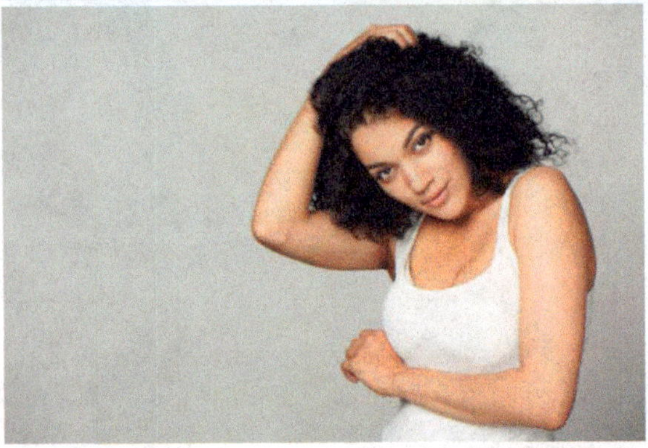

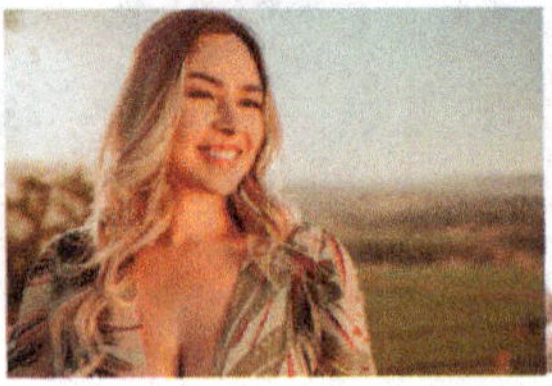

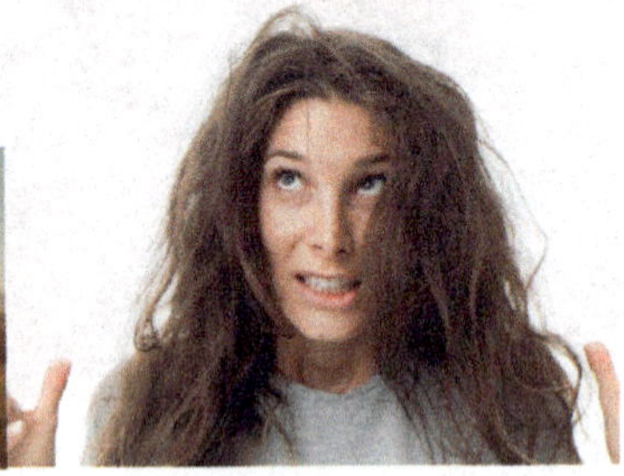

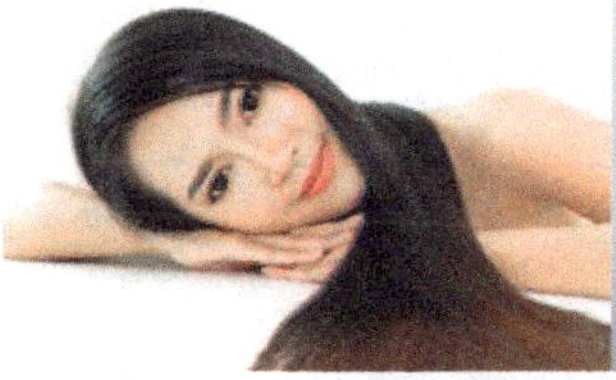
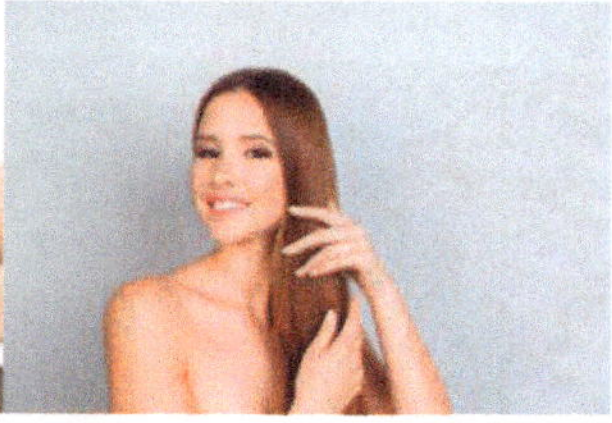
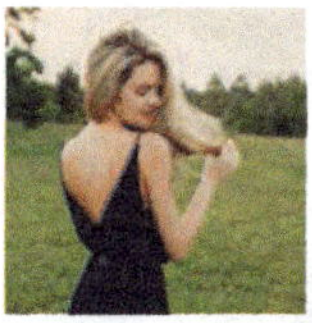
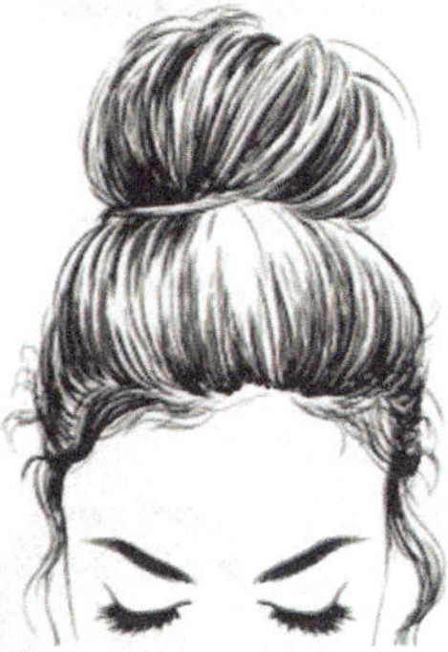
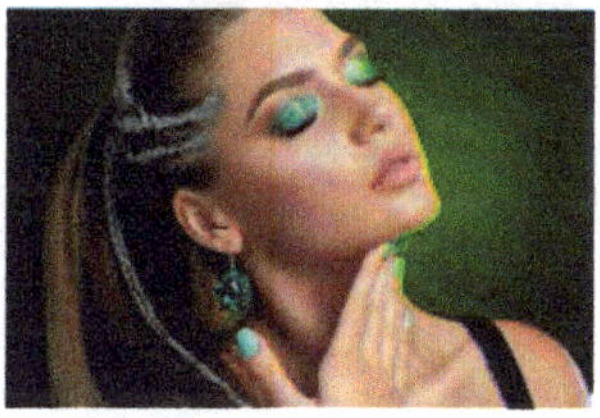
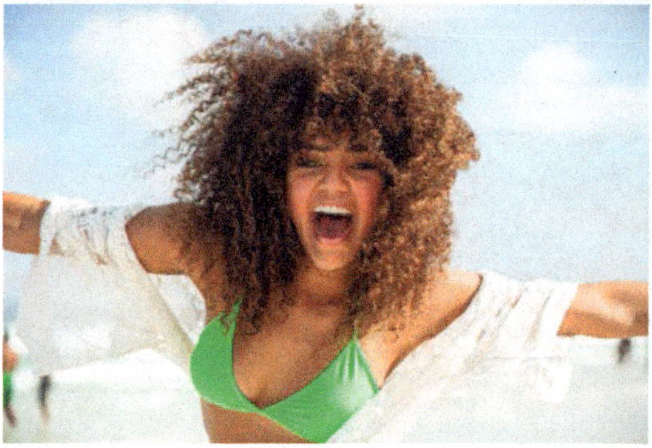

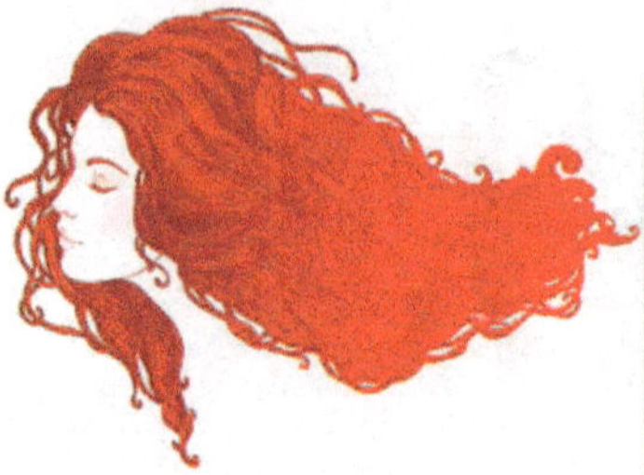
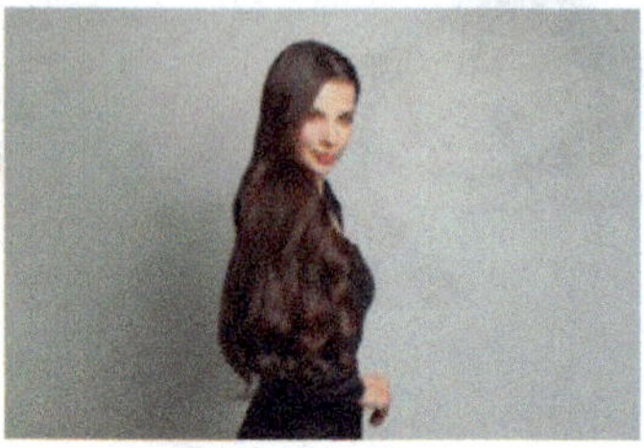
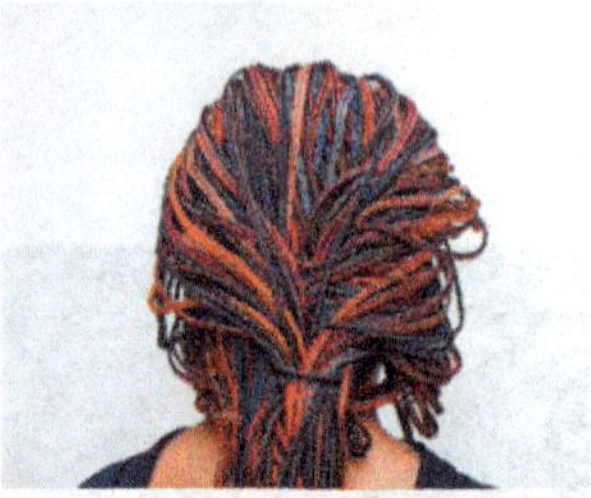
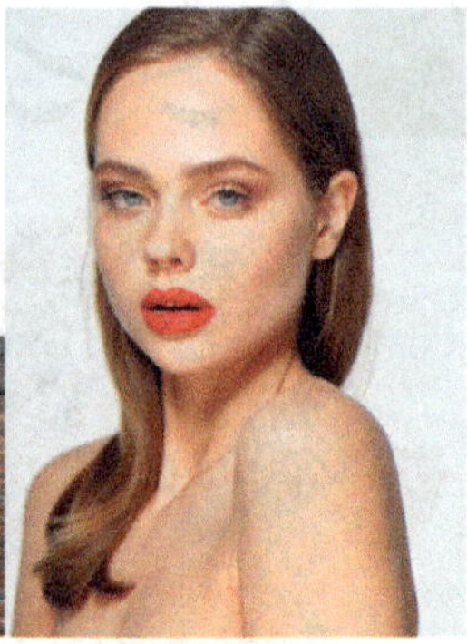
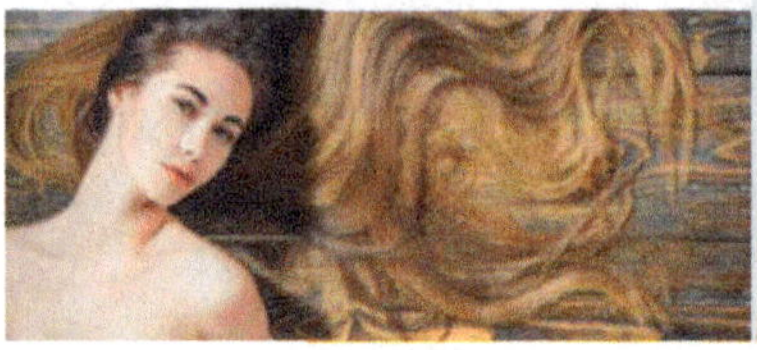

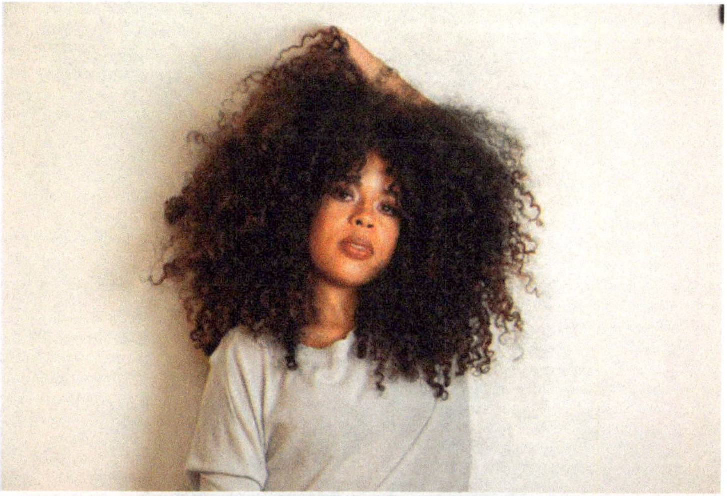

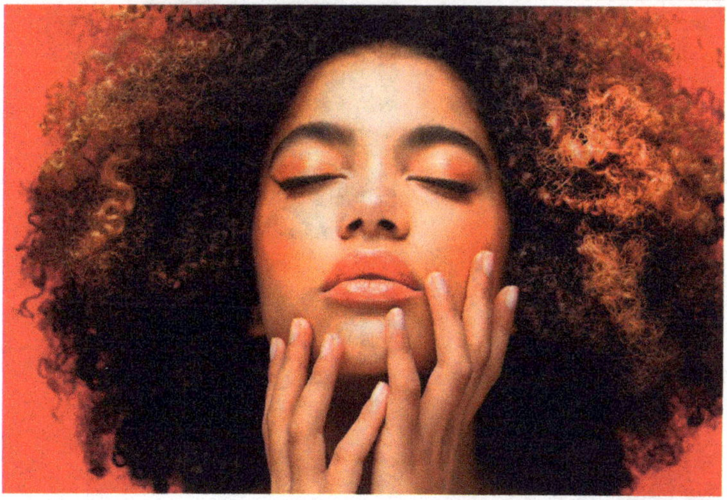

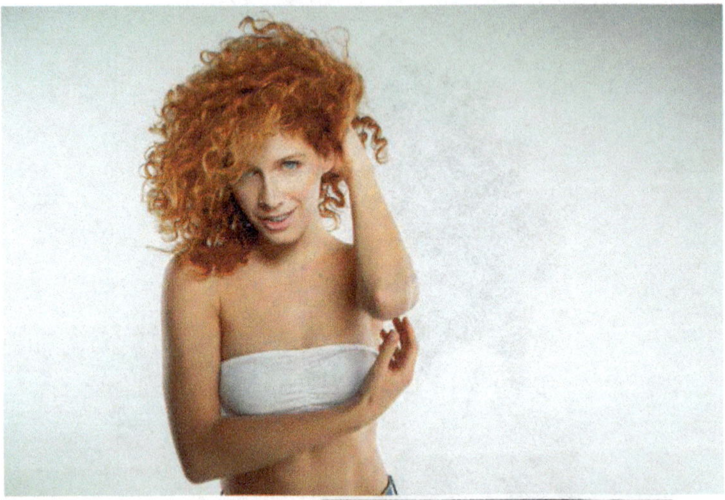
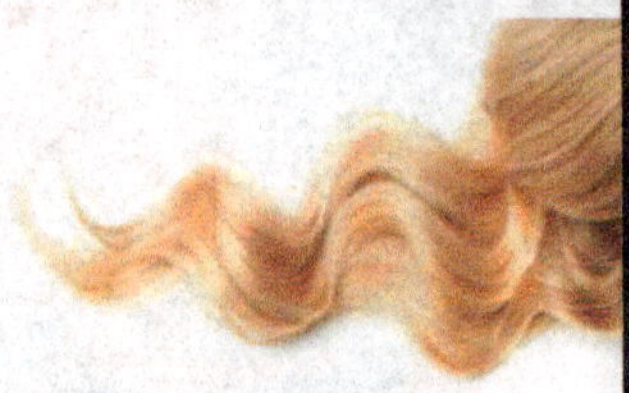
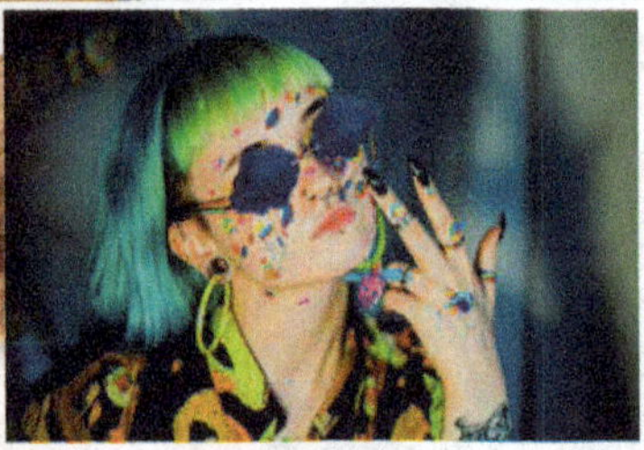
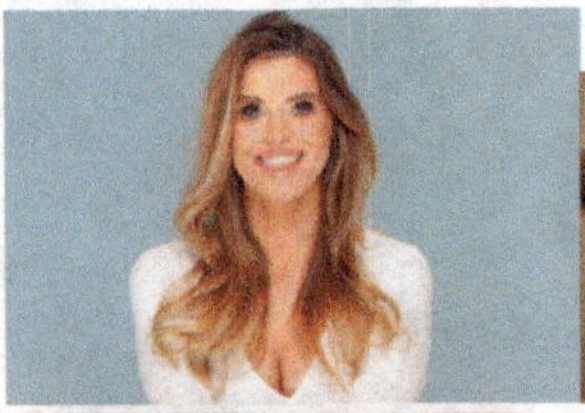
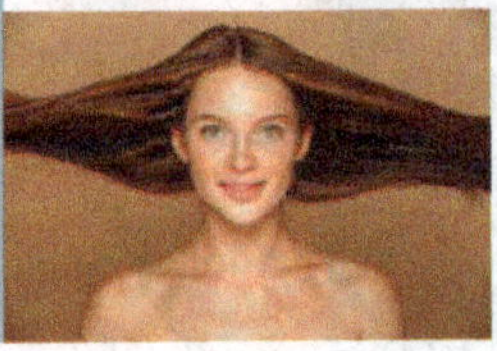
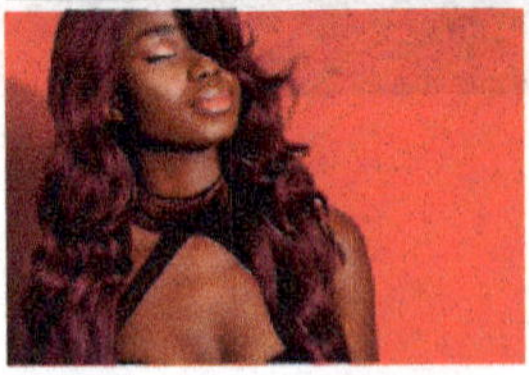

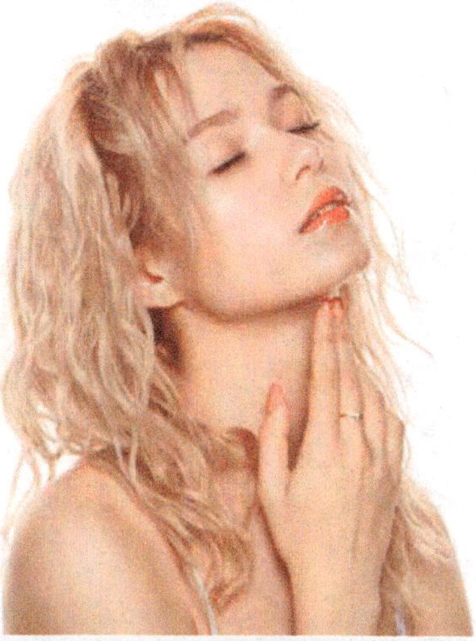

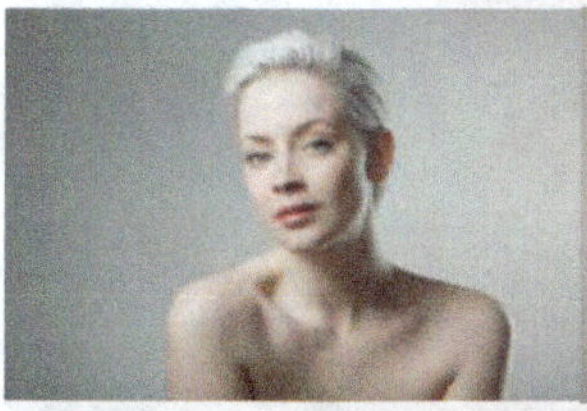

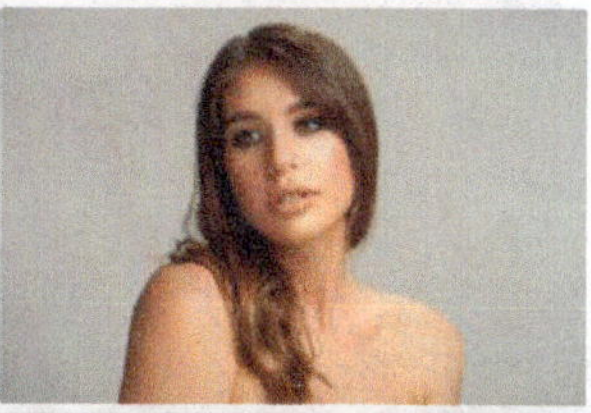

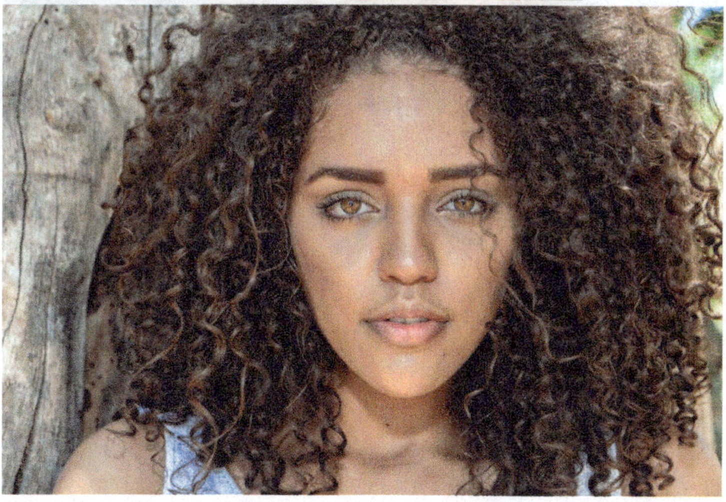

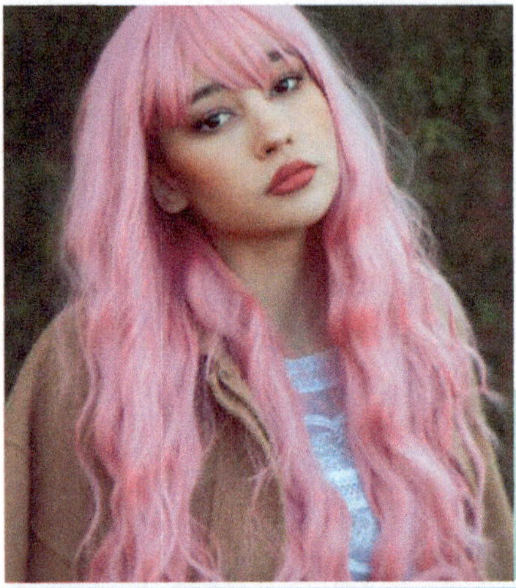

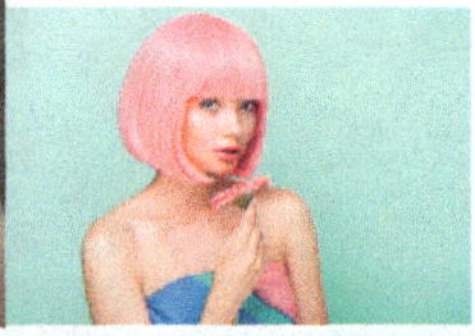

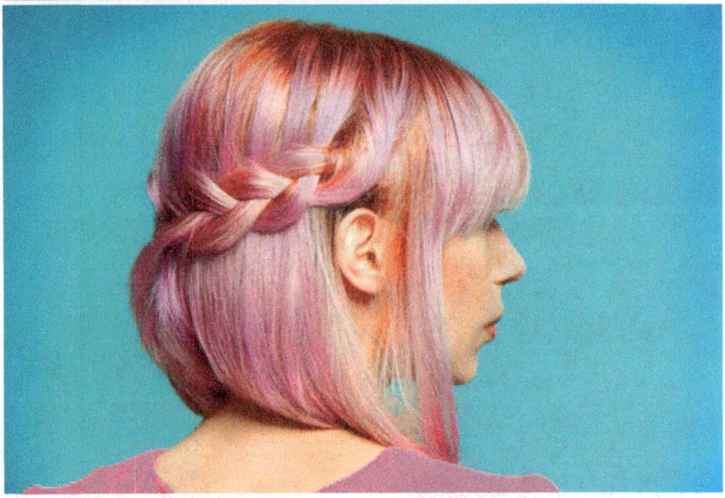

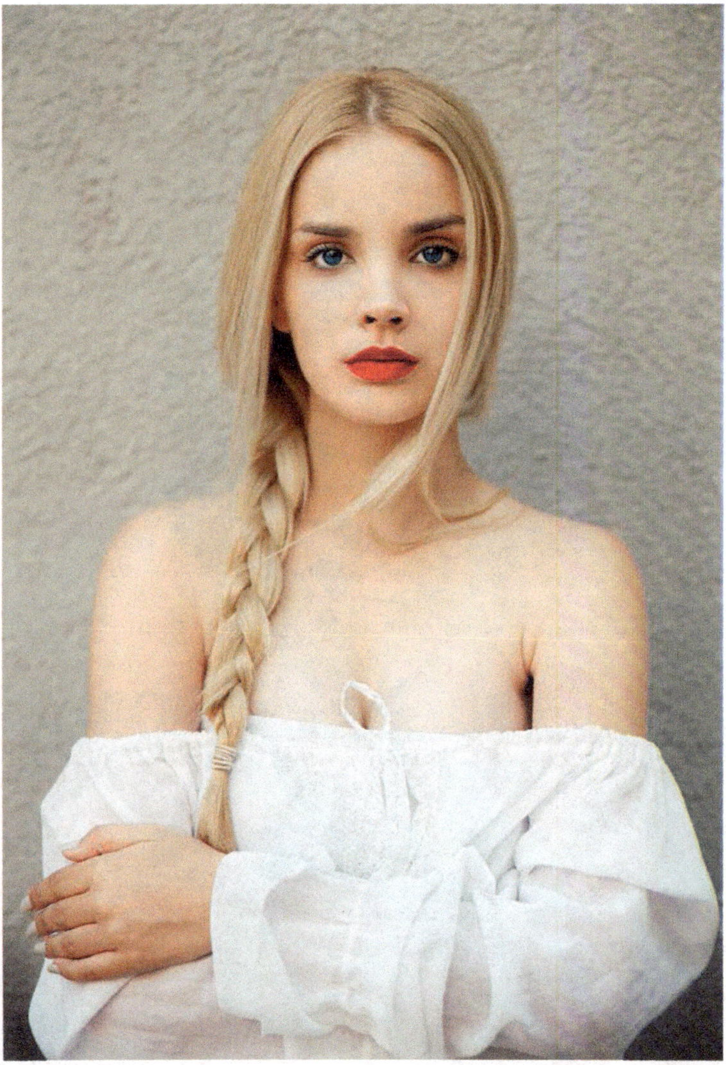

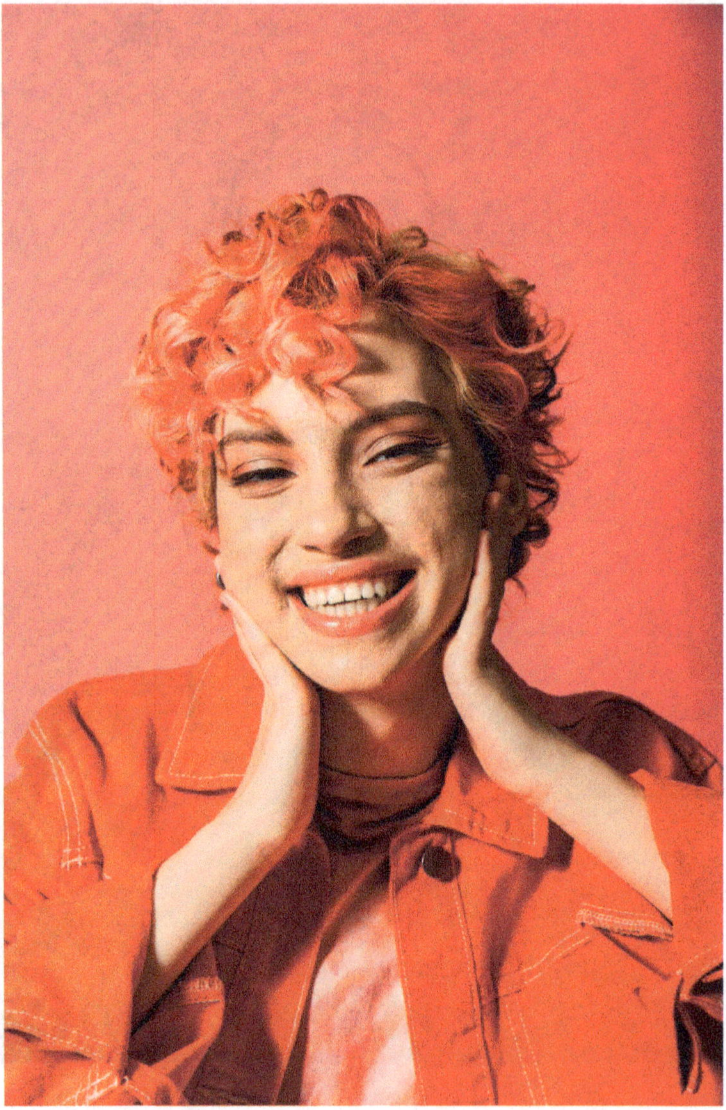

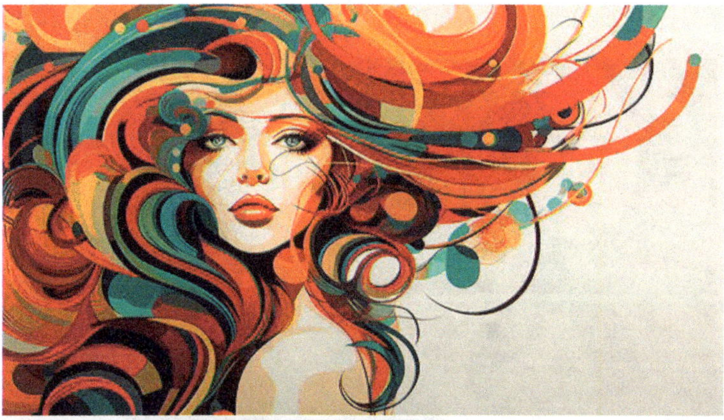

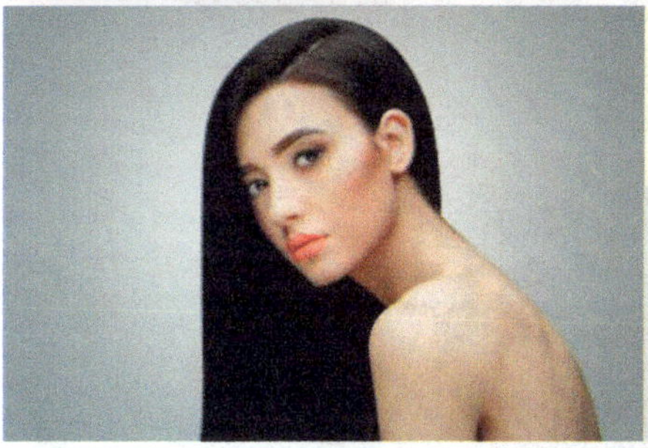

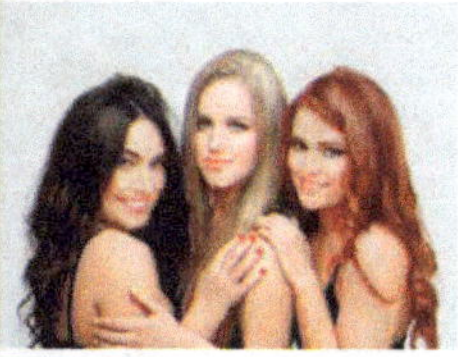

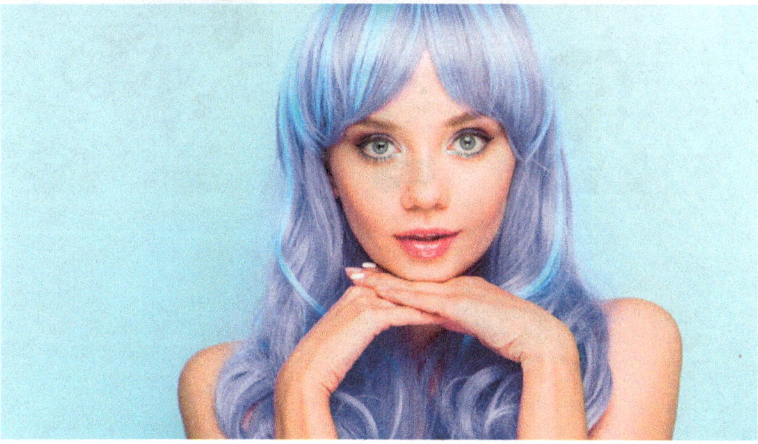

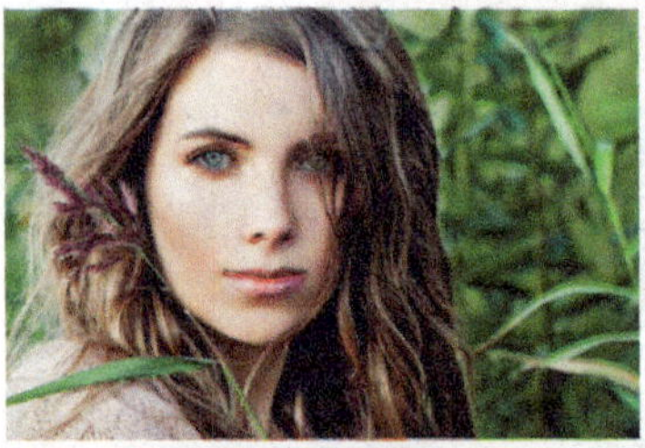
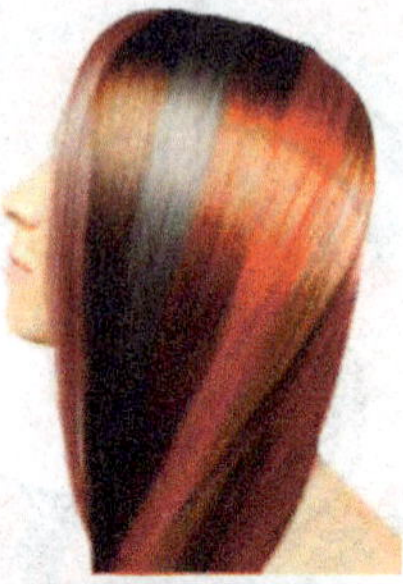
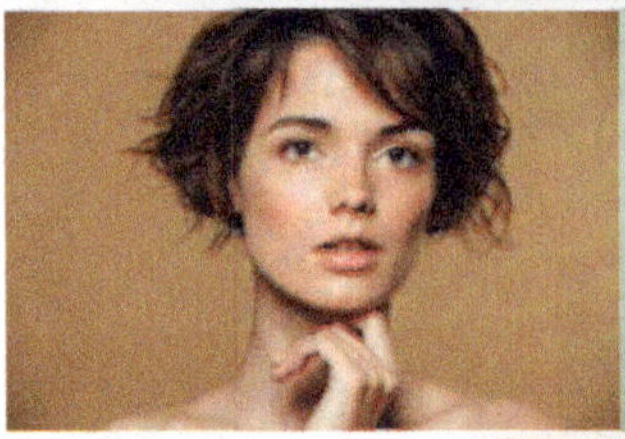
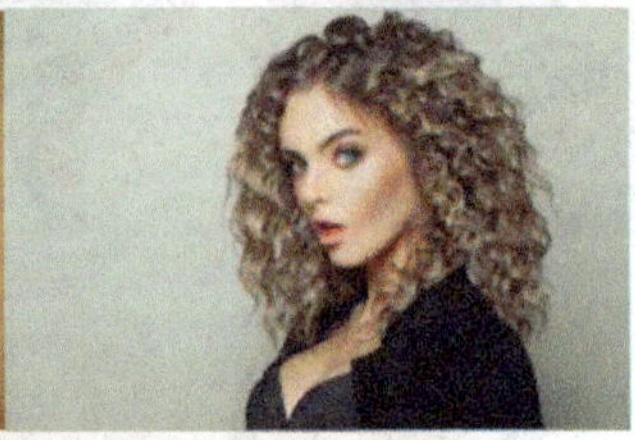
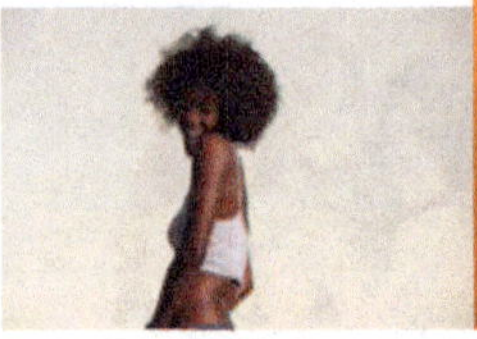
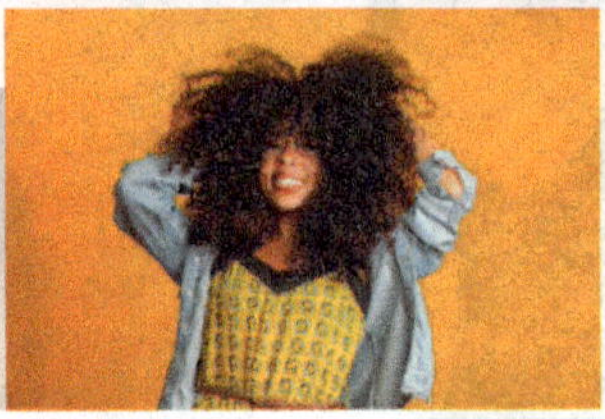
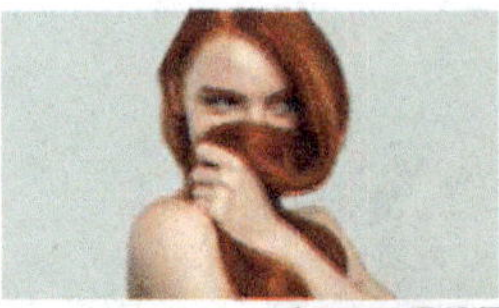
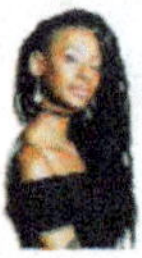
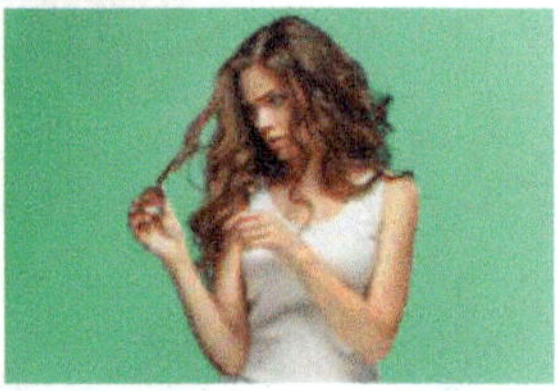
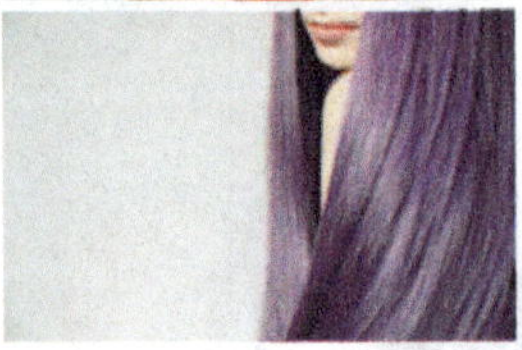

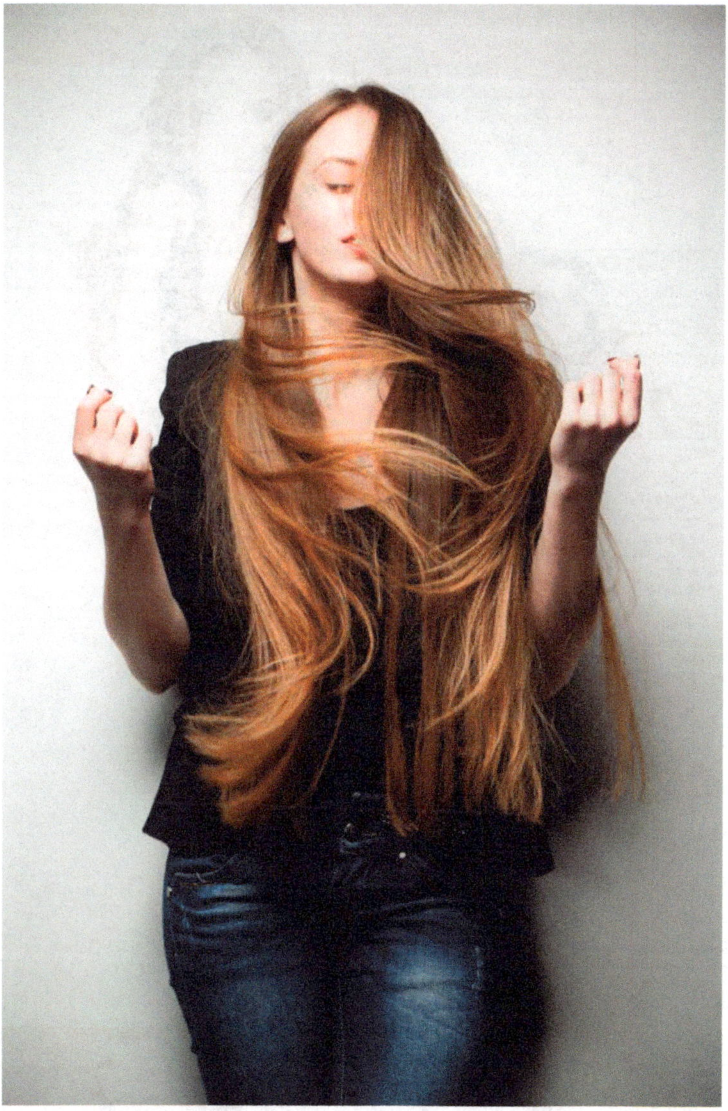

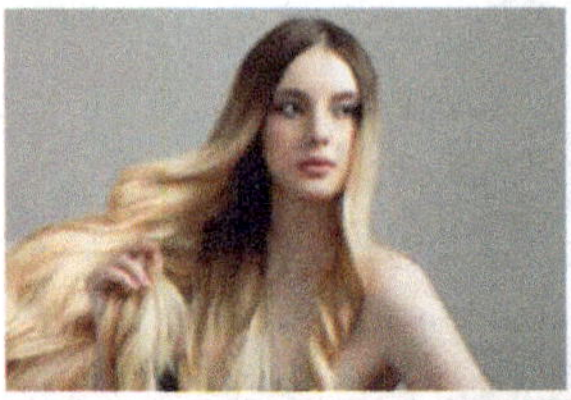
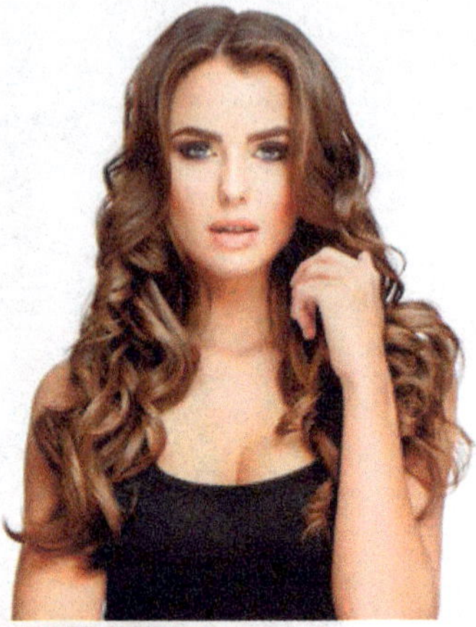
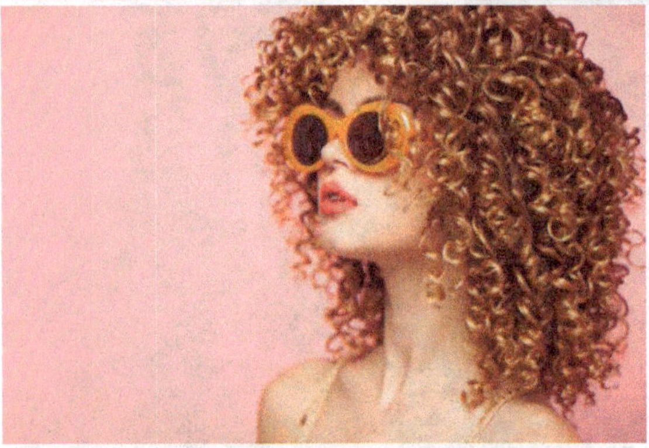
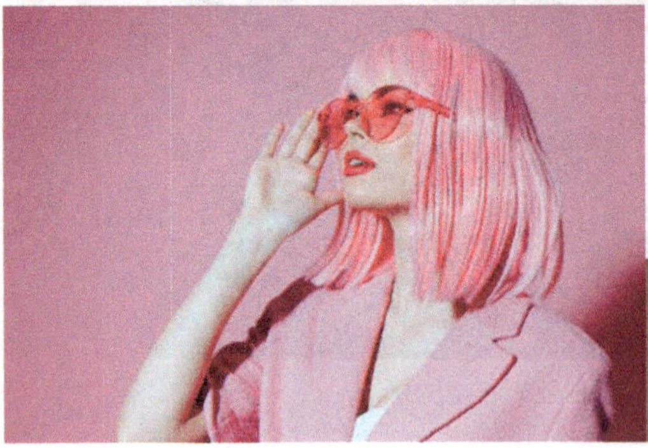
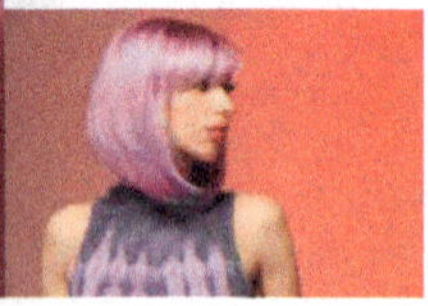

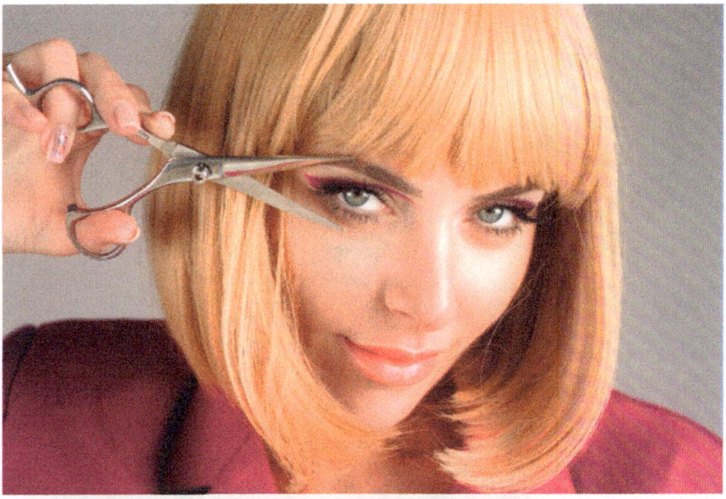

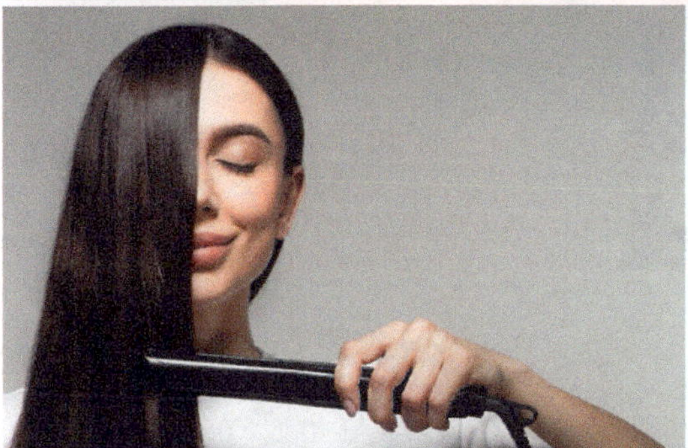

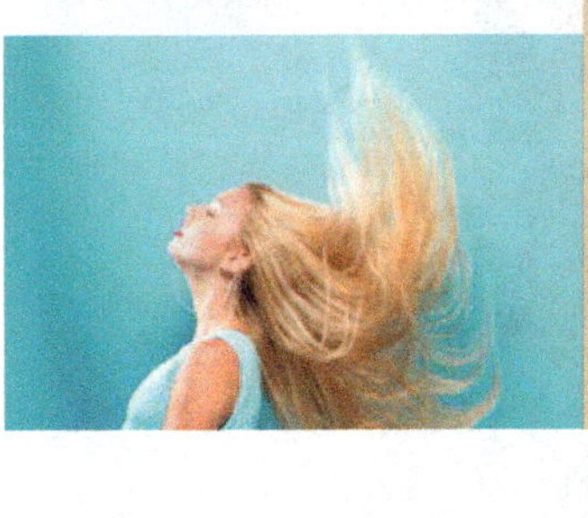

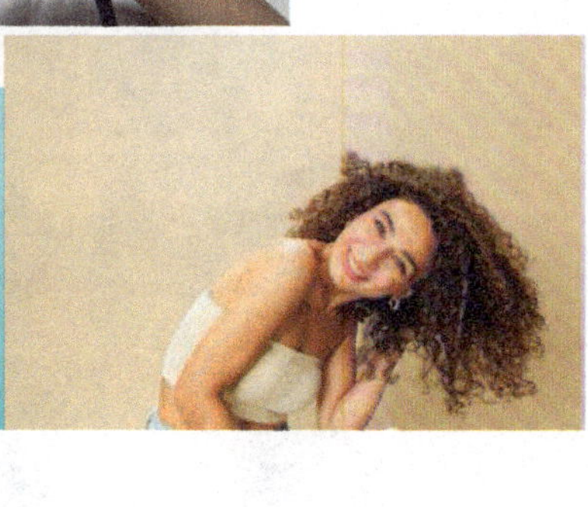

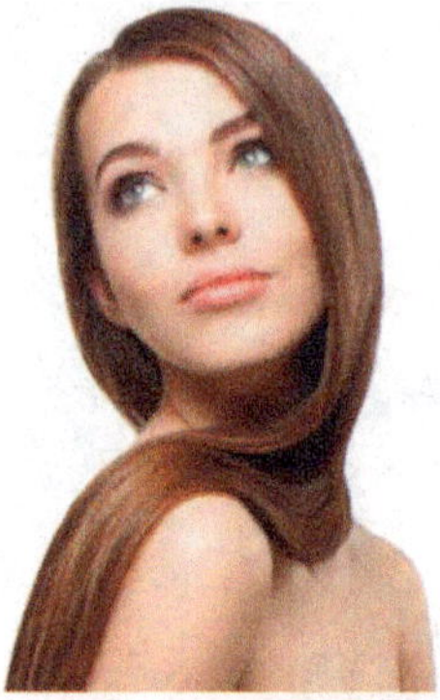

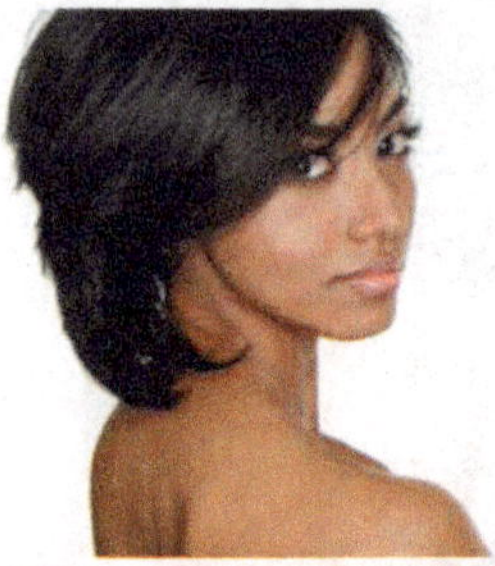

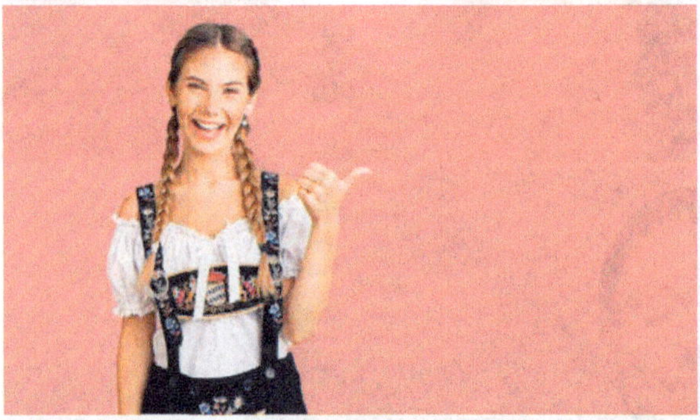

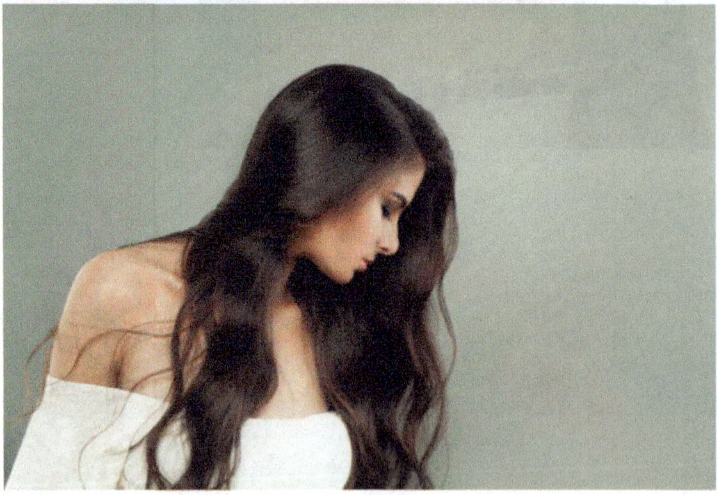

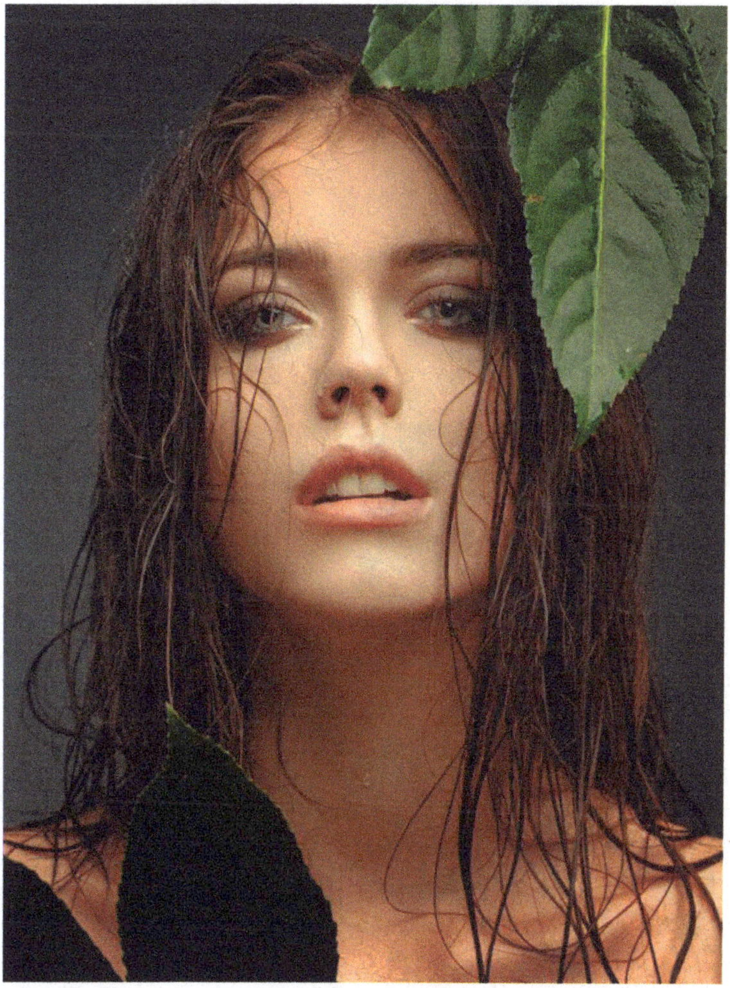

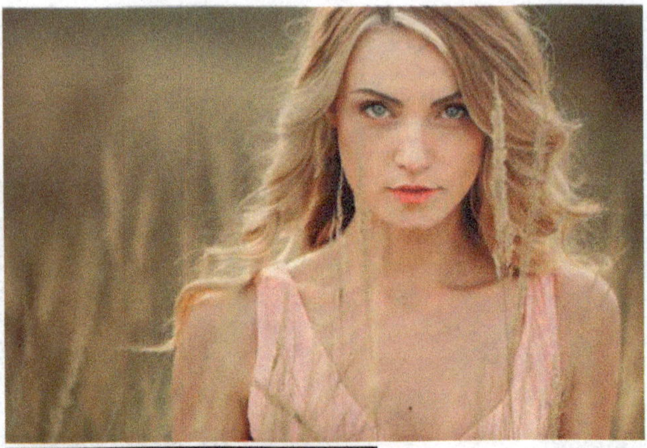

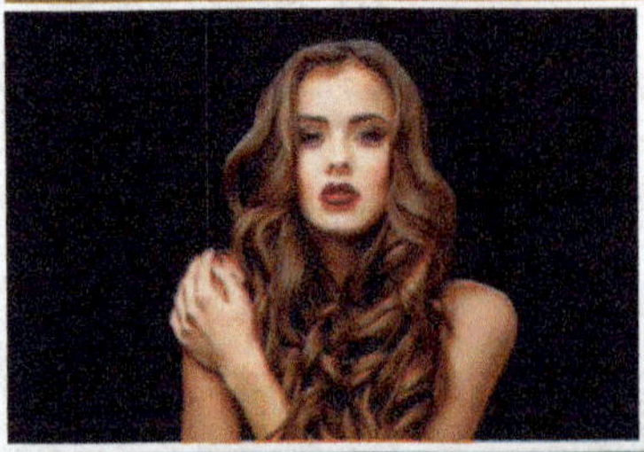

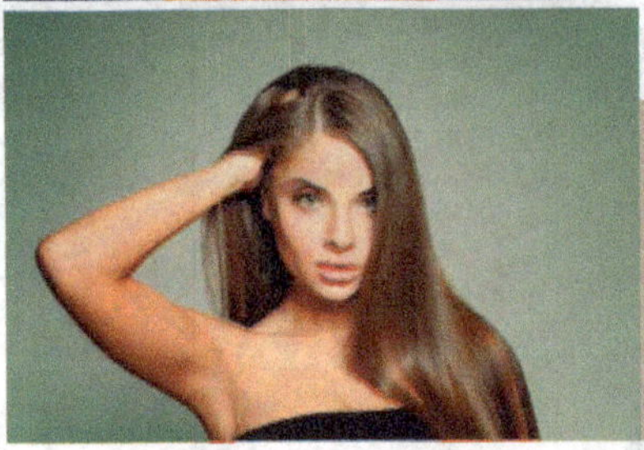

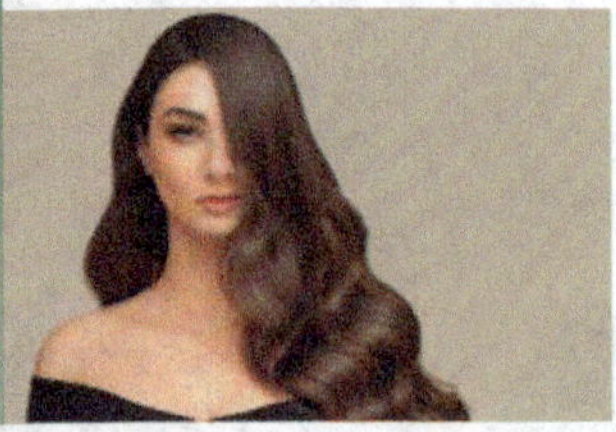

THE FOLLOWING PAGES ARE PAGES FROM THE BOOK "THE FEMALE FORM FROZEN" AND ARE COLLAGES THAT ARE ONE OF A KIND IN ORIGINAL PAPER AND COLOR FOR SALE FOR 1500 DOLLARS A PIECE

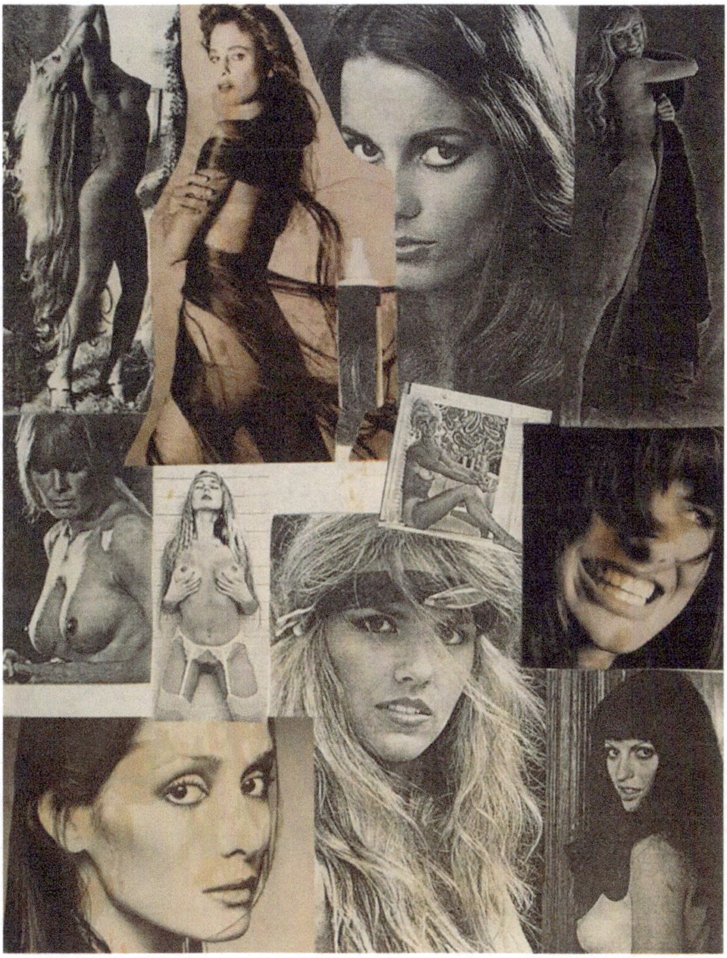

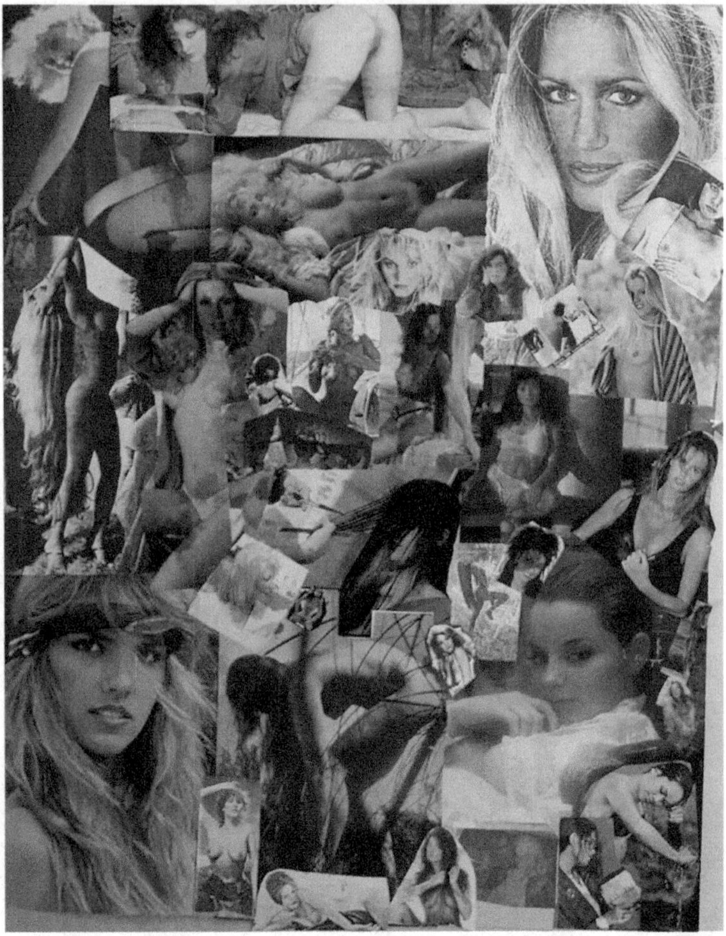

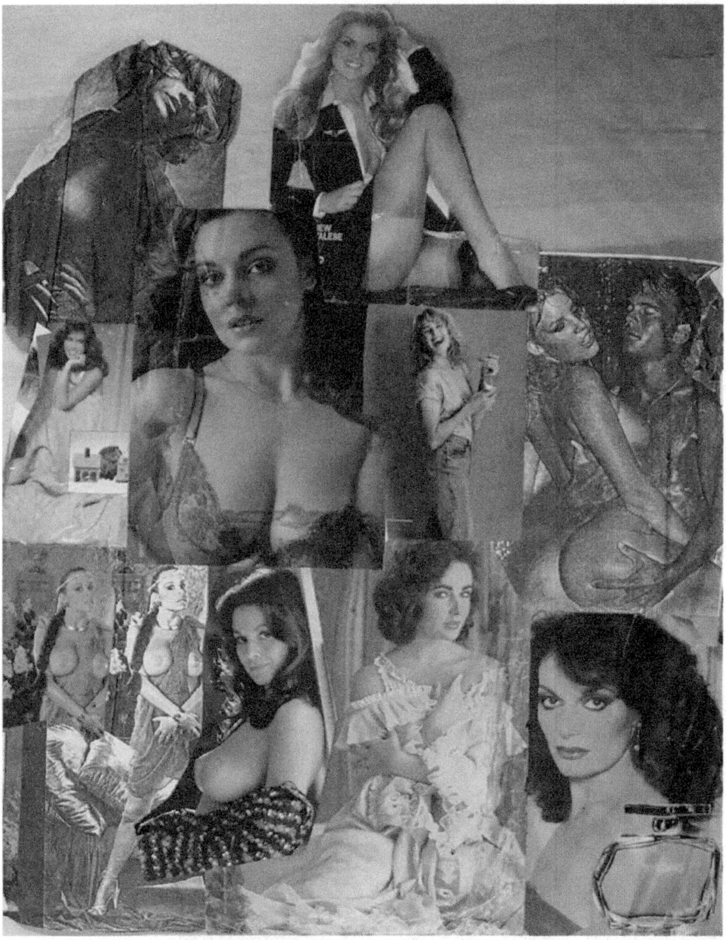

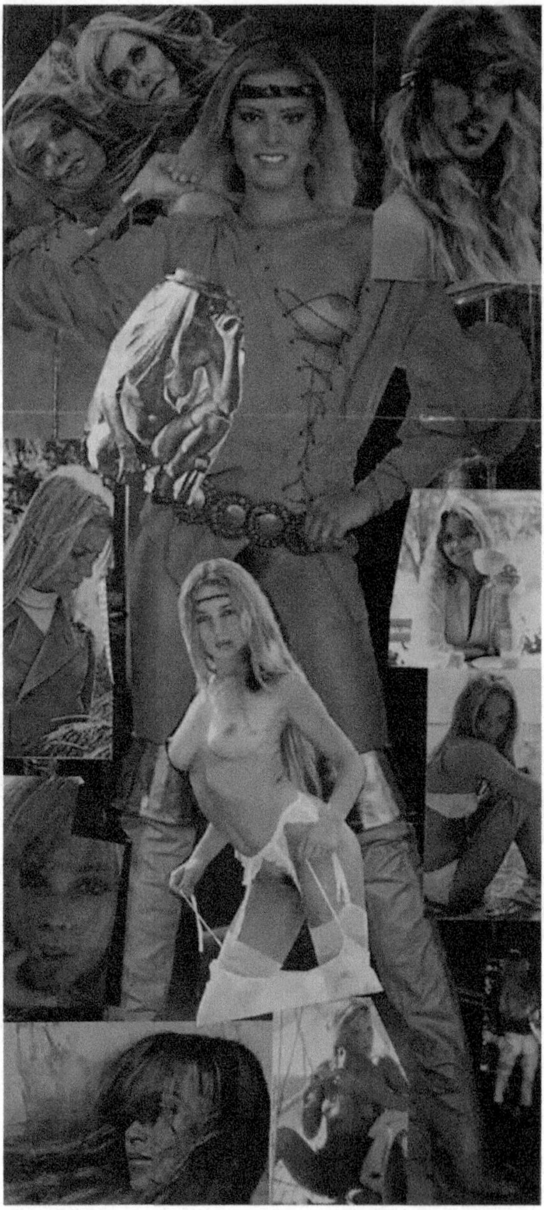

TORSO

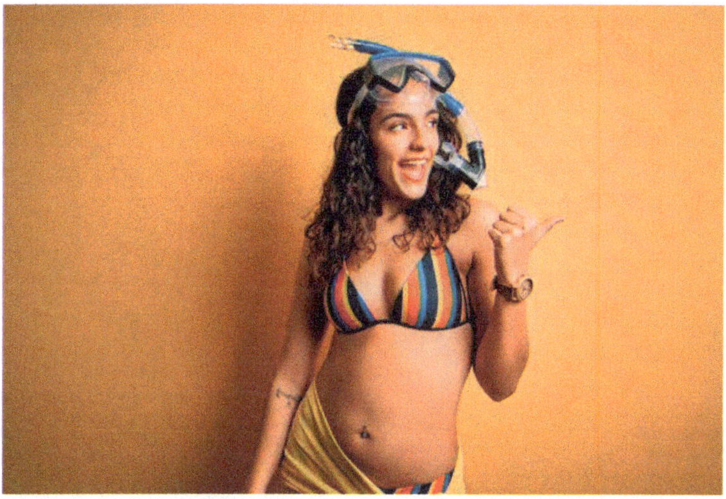

Made in the USA
Monee, IL
03 November 2026

67385839R00085